INTERFACE

INTERFACE

Branden Hookway

THE MIT PRESS
CAMBRIDGE, MASSACHUSETTS
LONDON, ENGLAND

Cover art taken from "From Flutter to Tumble: Inertial Drag and Froude Similarity in Falling Paper," by Andrew Belmonte, Hagai Eisenberg, and Elisha Moses, *Physical Review Letters* 81 (1998): 345–348, figure 1. Used with permission of Andrew Belmonte.

MIT Press books may be purchased at special quantity discounts for business or sales promotional use. For information, please email special_sales@mitpress.mit.edu.

This book was set in PF Din Pro by The MIT Press. Printed and bound in Spain.

Hookway, Branden.
Interface / Branden Hookway.
    pages cm
Includes bibliographical references and index.
ISBN 978-0-262-52550-3 (pbk. : alk. paper) 1. Technology—Philosophy. 2. Interfaces (Physical sciences). 3. Human-machine systems—Philosophy. I. Title.
T14.H64 2014
601—dc23
                                2013022300

10 9 8 7 6 5 4 3 2 1

TO MARIA

CONTENTS

PREFACE

This study views what I term "the interface" as a relation with technology rather than as a technology in itself. In this relation the interface describes a boundary condition that is at the same time encountered and worked through toward some specific end. In a way, my experience of coming upon the interface as a subject is analogous to this process. What started out as an introduction to a dissertation on the airplane cockpit as a paradigmatic twentieth-century environment became a separate project, as I found it necessary to work through a theoretical exploration of the interface in order to address what is at stake with the cockpit. The cockpit is at once a space of inhabitation, an ergonomics of use, an assemblage of mechanical articulations directed toward control surfaces and the materiality of air flow, and a threshold between human and machine whose mediation is expressed in a trajectory of flight. It encompasses a multiplicity of derivations, testing apparatuses, and simulations. As such, the cockpit has remained an implicit challenge in the theory of the interface presented here, which needed to account for the range of its instantiations, behaviors, and transformations.

Today the interface is at once ubiquitous and hidden to view. It is both the bottleneck through which all human relations to and through technology must pass, and a productive moment of encounter embedded and obscured within the use of technology. It is a disputed zone, a site of contestation between human beings and machines as much as between the social and the material, the political and the technological. In staging and resolving this contestation, the interface both defines and elides difference; it at once separates classes and draws them together as a single augmented body. While the interface operates in space and time, and on occasion may be described as a site or an event, it also governs the production of sites and events; it describes the site or moment in which the full operation and apparatus of systems, networks, hierarchies, and material flows are distilled into concrete action.

The aim of this study is to provide a theoretical framework for the interface and to examine the implications it holds over life. Chapter 1, "The Subject of the Interface," positions the interface with respect to theories of subject formation, agency, power, and control, and within contexts that include the technological, the political, and the game. The subject here is shown as poised between the simulation and the real, between autonomy and control. Chapter 2, "The Forming of the Interface," finds the origin of the term *interface* in nineteenth-century fluid dynamics, particularly in the work of James Thomson and James Clerk Maxwell, and subsequently traces its migration to thermodynamics, information theory, and cybernetics. As the site upon which Maxwell's demon first appears, the interface is shown to have a particular relevance to complex, dynamic systems, within which it describes the possibilities of agency and governance. Chapter 3, "The Augmentation of the Interface," addresses notions of tacit or embodied intelligence as they relate to what has been called the human-machine system. Throughout, the figure of the subject is inseparably both receiver and active producer in processes of subjectification. The interface is the endgame of a technological lineage, an architecture-as-medium that stands in a relation both alien and intimate, vertiginous and orienting, to those who cross its thresholds and trace out promenades in its interior places.

ACKNOWLEDGMENTS

Much of this book was first drafted in 2010 while I was living in Ithaca, New York. While writing on the airplane cockpit for a Ph.D. dissertation, it became clear to me that I needed to develop a theory of the interface first. I am grateful to Beatriz Colomina, my advisor, whose support and encouragement were invaluable throughout my studies and work on this project. M. Christine Boyer, my second reader, made discerning comments on the manuscript that helped refine it at various points, and I am thankful as well for her support. I am grateful to Mark Wigley and Alexander R. Galloway for their generosity and expertise as external readers. Among those who inspired me during my time at the Princeton University School of Architecture, I would particularly like to thank Hal Foster, Jonathan Crary, Spyros Papapetros, and Sarah Whiting for their timely encouragement. I am grateful to have had the chance to present and refine this work in an academic setting and would especially like to thank Mohsen Mostafavi, Kent Kleinman, Robert Somol, Ben van Berkel, Michael Bell, Lily Chi, Mark Cruvellier, Iftikhar Dadi, Sheila Danko, Stan Allen, Lars Lerup, and Bruce Mau for supporting this work. I would also like to thank the Cornell University College of Architecture, Art, and Planning for research and publication support.

At the MIT Press, I am profoundly grateful to Roger Conover for taking on this project. His guidance and vision were critical in bringing the work to its current state. I am also grateful to Thomas Frick and Matthew Abbate for their care and precision in editing, Justin Kehoe for his help throughout the process, Margarita Encomienda for the thoughtful design of the book, and the anonymous readers of the manuscript for their useful comments.

I am very grateful to my family for their love and support. My deepest thanks go to Maria Park, whose faith in this project and insightful reading of the work in all stages helped greatly in shaping the flow of the book. Finally, I would like to thank Lucy and Joseph, our most constant and best companions throughout this journey.

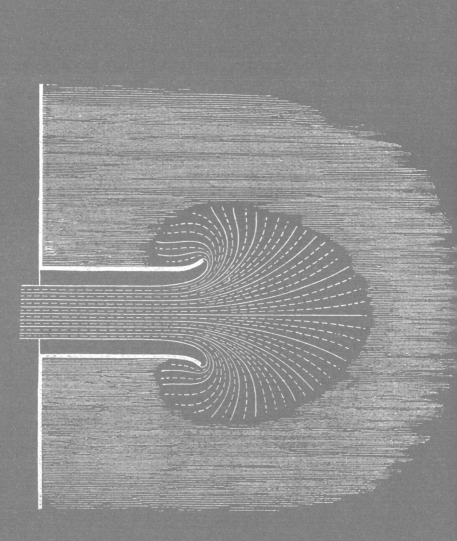

1

THE SUBJECT OF THE INTERFACE

/

The interface as form of relation

Inasmuch as the range of human experience and performance is more and more defined and conditioned through the forces of technological development, the interface holds a familiar albeit indeterminate and even spectral presence. For while the interface might seem to be a form of technology, it is more properly a form of relating to technology, and so constitutes a relation that is already given, to be composed of the combined activities of human and machine. The interface precedes the purely technological, just as one encounters a mirror image before the mirror itself. Likewise, the interface describes the ways in which humanness is implicated in its relation with technology. For even at the moment human and machine come into contact, their encounter has already been subject to a mediation. Both the actions performed upon the interface and the agency of their performance are to a critical extent already anticipated.

Nonetheless, it is the interface that most actively determines the human relation to technology and delimits the boundaries that define human and machine. Increasingly the interface constitutes the gateway through which the reservoir of human agency and experience is situated with respect to all that stands outside of it, whether technological, material, social, economic, or political. It is more and more unavoidably the means of representing that which is otherwise unrepresentable, or of knowing that which is otherwise unknowable. If the interface is now ubiquitous and pervasive, it is so with respect to a proliferation of ever more complex devices and networks. If it is indeterminate and elusive,

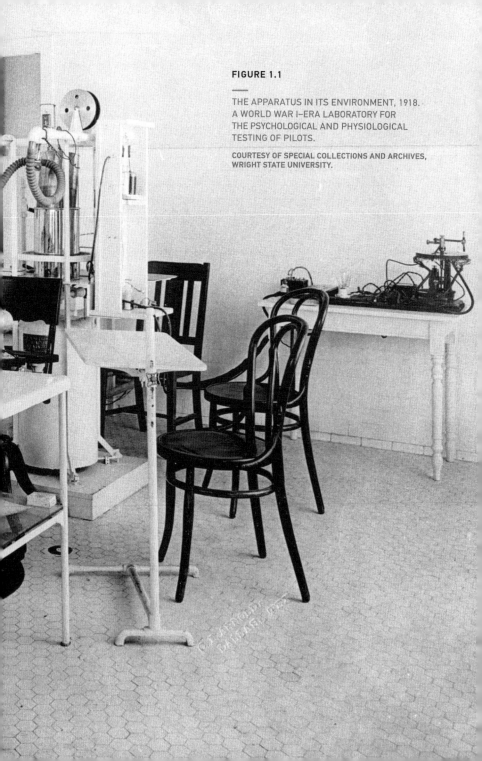

FIGURE 1.1

THE APPARATUS IN ITS ENVIRONMENT, 1918.
A WORLD WAR I–ERA LABORATORY FOR
THE PSYCHOLOGICAL AND PHYSIOLOGICAL
TESTING OF PILOTS.

COURTESY OF SPECIAL COLLECTIONS AND ARCHIVES,
WRIGHT STATE UNIVERSITY.

it is so in that it channels the activities under its influence toward a resolution within a common protocol, while at the same time opening up new vistas and capabilities to a now-augmented human sensorium.

The interface is defined here as a kind of theoretical construct whose essential characteristics and operations are common to each of its various realized instantiations. Specifically, the interface is treated here as a *form of relation*. This is to say that what is most essential to a description of the interface lies not in the qualities of an entity or in lineages of devices or technologies, but rather in the qualities of relation between entities. Such a relation possesses its own qualities and characteristics that are attendant on but otherwise independent of the entities brought into relation; the persistence of this relation in time and space is such that it may be described as possessing a kind of form. A preliminary definition of *interface* might then be as follows: the interface is a form of relation that obtains between two or more distinct entities, conditions, or states such that it only comes into being as these distinct entities enter into an active relation with one another; such that it actively maintains, polices, and draws on the separation that renders these entities as distinct at the same time as it selectively allows a transmission or communication of force or information from one entity to the other; and such that its overall activity brings about the production of a unified condition or system that is mutually defined through the regulated and specified interrelations of these distinct entities. Or again: the interface is that form of relation which is defined by the simultaneity and inseparability of its processes of separation and augmentation, of maintaining distinction while at the same time eliding it in the production of a mutualism that may be viewed as an entity in its own right, with its own characteristics and behaviors that cannot be reduced to those of its constituent elements.

The interface is defined in its coupling of the processes of holding apart and drawing together, of confining and opening up, of disciplining and enabling, of excluding and including. The separation maintained by the interface between distinct entities or states is also the basis of the unity it produces from those entities or states. While the constituent entities and processes of the interface may be examined individually, such analysis yields only a partial view of the interface and addresses only aspects or derivations of its full functioning. Such derivations of the interface include the surface, the test, and the simulation. The theory of the interface presented here investigates the interface both in part and in full, including the processes by which the interface comes into being, the behaviors and activities that it both draws upon and produces, and the status it ascribes to the discrete elements it brings into relation and the mutually directed entity or system that is the result of its operations. In this analysis, the interface entails implications for notions of control and intelligence as well as regarding those entities that are both its constituents and its products. These include the *system* and, perhaps most relevant to this study's focus on the human relation to technology, the *subject* and its production through processes of subjectification. The subject of the interface finds as its counterpart the user of the interface, just as the user's learning or mastery of the interface is at the same time a kind of subjectification. That the user of the interface is also its subject follows the notion of the interface as that which at once separates and draws together in augmentation. Likewise, *agency*, or the will and means to action, is a capacity at once mediated by and produced upon the interface.

5

The human-machine interface is neither the first interface nor the only type of interface that may be defined as a form of relation. The concept of the interface was developed for use in the field of fluid dynamics. Fluidity provides a powerful metaphor for the operation of the interface, as well as for associated processes of mediation and control. To engage an interface is also to become a constituent element within a kind of fluidity. Likewise, subjectification may be described as a process of becoming fluid.

The interface is a liminal or threshold condition that both delimits the space for a kind of inhabitation and opens up otherwise unavailable phenomena, conditions, situations, and territories for exploration, use, participation, and exploitation. Often the territories it opens up constitute in themselves further threshold conditions. This reflects what may be taken as axiomatic: that the interface is at every stage of its operation concerned with the liminal. Not only does the interface constitute in itself a threshold condition, but it also operates through the seeking out, identification, and development of thresholds of various kinds. These thresholds are guarded, regulated, and maintained in place by the interface both in its internal organization and in the relation or effect it produces with respect to the externality with which it interfaces. The relation of an interface to its external condition, a relation that is the primary product of its operation, may be described as control. Insofar as the interface serves as a locus and condition of control, control could also be said to pertain to the liminal, in that it describes a way of operating upon and through threshold conditions; this is to say that, at least in relation to the interface, control proceeds *a limine*, or out of a threshold. It is axiomatic of control as well, then, that it both occurs upon a threshold and proceeds from a threshold; control may even be said to define the threshold to the extent that it seeks out those moments, or tipping points, at the onset of a transition from which a difference may be most easily effected. To the extent that the identification of difference is essential to the operation of the interface, the interface is aligned with the test; and to the extent that the interface occupies the threshold that governs the change from one state to another, the interface may be said to possess a tendency to come into being, operate within, and express its character with reference to the transformative or transitional.

This is borne out in the history of the human-machine interface from the early twentieth century to now. During this period the interface has become a prevalent means of testing and simulation, has served as a testing ground for transformations in self-identity, and has been the site from which complex technological processes are governed, from the control of machinery to the design of environments to the modeling of complex physical processes. In each of these settings, and whether as a general theoretical construct or within a specific instantiation, the interface carries with it a third major tendency, along with the identification of differences and the facilitation of transformations; this is a tendency toward a seeming transparency and disappearance, even as it is undoubtedly a condition that demands to be worked through. While promising an

6

illusory effortlessness and seamlessness in its provision of an augmentation, the interface nonetheless requires an extraction of work and for this work a cost must be paid. This cost is extracted both in terms of energy and in the confinement and channeling of these energies into a form compatible with the interface, even as the cost of working through the interface is hidden from the perspective of its having been worked through. In its occupation of the threshold, the interface is both the conduit through the threshold and the judge sitting upon the threshold to determine what may pass through and the manner of its passing. Both of these aspects of the interface constitute a kind of friction upon the threshold that requires work or the exertion of energy to overcome. What occurs within the interface, the kind of relating across a threshold that is often described as interaction or interactivity, may also be described as a transaction, in the sense of a cost being extracted and compensation being given in exchange. This transaction also reflects the reconciliation of the interface as a space that is both inhabited and worked through; here the transaction is a confinement endured for the granting of an enhancement.

Between faces and facing between

The etymology of *interface*, a word first used in the description of fluid behavior, suggests how the interface may be opened up to theoretical description even as it resists such description. The prefix *inter-* connotes relations that take place within an already bounded field, whether spatial or temporal. It pertains to an inward orientation, an interiority. As an interiority of relations, *inter-* encompasses relations that may occur between, among, or amid elements insofar as they are given as bounded within the space of their relating, or of events insofar as they are bounded in time. *Inter-* holds its bounded condition as already given, as a priori to the relations it describes. It does not exclude that which is exterior to it, since it has already been separated out as an interior. This reading of *inter-* would suggest an interface that does not define its bounding entities but is rather defined by them. The interface thus would be an interior condition, whose activity and influence is constrained within the boundaries given by its defining entities. If used as a form of communication between these entities, the role of the interface would be limited to the translation or transmission of that which its bounding entities project into it. While the specific means of this communication belong to the interface, the interface would otherwise always refer back to its bounding entities. Its influence would not extend into the bounding entities that confine it, but would rather be constrained to the relations that occur between them. The interface would be defined according to its betweenness, its amongness, its duration-within.

Against this reading of the interface as an interior condition, the etymology of *face* points toward an outward orientation and an exteriority. *Face* is derived from the Latin *facies*, meaning like the English *face* a visage or countenance, as well as an appearance, character, form, or figure; *facies* in turn is derived from the verb *facere*, meaning to act, make, form, do, cause or bring about. A face, then, is the aspect of a thing by which it presents itself. From *facere*, this is an active making of a presence, or a presencing. A face is not that by which a thing looks at itself, as into its interior; it is rather the focus of a relation of a thing to what is outside itself, to an exterior. In this way a face not only forms the outer boundary of a thing, but is also the means by which that thing may project itself forward and outside itself, and so by which it may enter into relation with something outside itself. The face of a thing is what is given as available for a reading; from its face one may determine the character or nature of a thing. As a verb, *to face* may broadly be said to have two meanings. First, *to face* is to give a thing the properties of possessing a face, such that it both becomes capable of projecting qualities and energies outside itself and is opened up as accessible to a kind of reading, just as a text is available to be read. This meaning of *to face* may be found in the concept of facing a building with marble or facing (making smooth) a block of stone to prepare it for use in building; in both cases *to face* is to produce a face through which an entity may present qualities outside itself in order to be read. Second, *to face* is to be oriented toward, or to confront with persistence and determination, as in an adversarial situation. Here, *to face* is not yet to enter into a relation, but rather to marshal energies from an interior toward an exterior. The face is the threshold for this marshaling; it is the site from which the qualities of an interior are translated into a communicative or combative form, so that they may be projected outward onto an exterior.

The combining of *inter-* and *face* makes of the interface the embodiment of a contradiction, which may be seen in two possible readings of the term. First, as "between faces," interface would suggest activities within a circumscribed field or an enclosure. Second, as "a facing between," interface would suggest a boundary or zone of encounter that actively extends into and conditions that which it separates. In combination, the interface is both an interiority confined by its bounding entities and a means of accessing, confronting, or projecting into an exteriority. It is defined by its bounding entities at the same time that it defines them. In encompassing interiority and exteriority, passivity and activity, the interface governs transformations from interior state to exterior relation, from inward to outward expression. Each successive state of such transformation belongs to the interface, as does the overall event of transformation itself. The interface, then, is at the same time "between faces" and "a facing between." Either reading may constitute a valid approach to the study of the interface, although both remain partial and provisional descriptions. The interface comes into being in the maintenance of its contradictions. It is only by maintaining these contradictory readings that the entire range of activity that may occur within and through interfaces may be addressed as belonging to a single theoretical concept.

One between-faces approach to the interface would be to treat it as if it were a closed system. The interface could then be characterized according to the bounding entities (or faces) that delimit it, and by the relations that take place within this delimited field. A human-machine interface, for example, would be fully bounded by the "faces" of human and machine. Its study would concern only the relations that take place between human and machine, and its operation would be delimited as acts of transaction and translation between these two entities. The view of the interface as an instrumental technology is such a between-faces approach. Here, the interface becomes a discrete object or apparatus

9

available for use, or a technical problem constrained within the criteria of its design and production. A standard definition of the human-computer interface—"the means of communication between a human user and a computer system, referring in particular to the use of input/output devices with supporting software"[1]—reflects this instrumentalist approach. Defined by an already-given accessibility to the designated methods and tools of a specialized discipline, the interface is posed as a design problem that aligns seamlessly with the technical means of its solution. Its bounding entities, human and machine, are treated as constants rather than variables themselves subject to the operations of the interface. While the instrumentalization of the interface is of use in analysis or design, just as one element of a highly complex open system may be singled out and viewed as a closed system, it provides only a partial view. To address what is at stake in the historical emergence of the interface, or the role of the interface as a cultural form, a countervailing reading of interface as a facing between—as an active and contested boundary condition—is needed.

At the same time, the reading of the interface as between faces reveals those aspects of its operation where it delimits, encloses, or produces an interiority. This is not only the interiority of the closed system, but also a form of interiority that defines the subjective experience of control interfaces, or of media of control. Here, the interface opens up a space of inhabitation, within which the use of a control interface and its exertion of control are reconciled within user experience. In this reconciliation, the operator of a control system projects agency through an interface, with the actual operations on the interface performed at a tacit or subliminal level of awareness with respect to the conscious exertion of control over an environment. Thus, a video gamer internalizes the use of controls to project an agency or selfhood into the world of a game, or the user of social media internalizes the protocols of the social network in adopting a social identity. These inhabitations are only ever partial and contingent; they remain fully reliant on the act and apparatus of projection even as they obscure that projection. The partial self that inhabits the game world is a kind of abstraction within the full operation of the interface, just as the closed system is an abstraction of the open system from which it is separated.

As a facing between, the interface is no longer defined by enclosure but rather actively faces that which it encounters. In this sense the interface defines its own interiority in exclusion of its bounding entities, and so possesses its own specific qualities and tendencies beyond those derived from its bounding entities. In possessing its own faces, the interface also possesses the agency by which it is capable of facing. This agency may be expressed as dynamic form, behavior, or intelligence. Here the interface is more than a means of communication between its bounding entities. It holds its own identity, from which it influences and defines the entities that stand in relationship to it as much as those entities influence and define the interface itself. In actively facing its bounding entities, the interface defines them according to the relation brought into being by that facing. The interface binds together its bounding entities and mobilizes them as constituent elements of a unified condition whose interiority is the interface. The interface is at the same time constitutive of this unified condition, in defining its interiority, and exterior to that condition, in that it continues to present a face to its constituent elements. In this way the interface describes a form of agency within a given condition that yet is not encompassed by that condition.

In exerting a form of agency at once interior and exterior to a condition, the interface also manifests the potential availability of that condition to control. The agency of the interface cannot yet be termed control, though it opens up the opportunity of control. The interface comes into being prior to control; while it does not necessarily entail control, it is the conduit of control, and control always takes place across an interface of some kind. Control recapitulates the binding together of entities by the interface, albeit in an implicit modeling of the interface as an exterior means of access to the interior processes of a condition. In this way control draws a loop diagram connecting the interior state of a condition and an exterior means of reference that models that condition. Yet the interface is not reducible to control, even as control implicitly seeks out the interface as underdeveloped territory to be explored and colonized.

With respect to control, the interface describes both a possibility and a limit, a capacity and complexity at once available to and beyond control. If control is also a means of understanding or knowing a condition, the interface stands at the limit of that knowing, in that it is the site from which that condition comes into being. It is as well the site from which the entities that are constitutive of that condition are defined as both active and acted upon, just as the interface defines its bounding entities according to the unified condition or mutual activity that it brings into being.

As a zone of encounter between entities, the interface is at once between faces and a facing between, just as it is at once passive and active. It comes into being between faces, constituting the site of encounter between two or more entities as they enter into relation; as much as this relation produces mutually determined activity, the interface operates as a facing between to bind together the actions and reactions of each entity in the production of an overall act. Likewise the interface is at once passive in that it only comes into being when energy is directed into and through it, and active in that it captures that energy as its own, drawing energies from one entity to channel it into another in the production of a mutual activity that only it can fully describe. To return to the human-computer interface, the interface is not only defined by but also actively defines what is human and what is machine. In this mutual defining, which is also both a communication and a contestation, the interface operates as an essentially unbounded condition—one that continually tests and redefines its own boundaries as it comes to face with the entities that face it.

The interface and the surface

As a boundary condition that comes into being through the active relation of two or more distinct entities or conditions, the interface may be distinguished from the *surface*. The *sur-face*, as a facing above or upon (*sur-*) a given thing, refers first of all back to the thing it surfaces, rather than to a relation between two or more things. A surface exists primarily as an aspect of that which it surfaces, before it can be said to perform any additional functions or hold any other characteristics that relate to its facing. Thus even in its mathematical usage a surface, as a topological manifold, refers back to the space within which it is generated. As an aspect of a thing that may open it up to a condition of relating even though it does not yet embody a relation, the surface is that which allows a thing to be regarded as an object. With respect to that regarding, the surface may still be said to possess its own distinction as an aspect of the thing that it surfaces, in the particular way that it refers back to that thing. This is just as a topology, screen, or landscape demands some form of reading to ascertain the properties of that which it surfaces and which brings it into being. Here, a surface could be viewed as an embodiment of the facing of the thing toward an exteriority, and so as a kind of culmination of the externally directed energies of that thing, as much as it could be seen as the means of concealing the interiority of a thing from external view. On one hand, then, the surface is the means by which a thing expresses itself and the means by which it may be read, while on the other hand the surface is that which must be penetrated or seen through in order to uncover those essential properties or traits of the thing that may be hidden or remain unexpressed in the formulation of what is expressed.

When viewed as a problem or an analytical method, the surface focuses attention on the ways in which the thing that is surfaced is referenced by that surface, both in the formation of that surface and in the ways in which the identity of that thing is bound up or reflected by its surface. If a surface is given as something that may be read or otherwise available to some form of testing, then setting the surface as a problem begs the use of methodologies by which the surface may be interrogated both on its own terms as a surface and in its means of referring to the thing that it surfaces, whether as an expression, a topology, a signifier, and so on. Such methodologies could be characterized as allowing a close reading of the surface. They may be analytic, hermeneutic, deconstructionist, genealogical, et cetera. The interface may be distinguished from the surface in that it does not primarily refer back to a thing or

13

condition but rather to a relation between things or conditions, or to a condition as it is produced by a relation. The interface as a problem does not primarily bear on the characteristics or properties of the entities it interfaces, though it may do so secondarily. Rather, the problem of the interface bears on what must take place in the drawing together of entities into a relation, and to the combined or synthesized behavior of those entities within that relation.

If the surface may be seen as the culmination, expression, or concealment of a thing, and so in varying ways the means by which a thing may be made available for theorization or some form of reading, then the interface may likewise be seen as the culmination, expression, or concealment of an active relation between things. What the theorization of the interface reveals is not the properties or essence of a thing but rather the interplay, within a relation, in the shaping of a mutually generated behavior or action. While the surface and the interface may each involve in their expression both a culminating and a concealing, what is meant by culminating and concealing differs in that the surface refers back to a thing and expresses the properties of that thing, while the interface refers back to a relation between things and expresses an action. An analysis beginning from the surface privileges the question of what a thing is or what its properties might be, while one beginning from the interface privileges the question of how a relation may come into being and how it may produce behaviors or actions. A surface presents a form, while an interface performs a shaping. Though it may produce a trajectory or even a surface, the interface nonetheless remains resistant or hidden to formal analysis. For example, to be illusory with respect to a surface is to refer to its role in masking some aspect or property of a thing, so that the surface brings about the illusory disappearance of that aspect or property. With respect to the interface, this illusory disappearance is performed as part of an active relation, and takes place as the concealing of constituent activities within the production of an overall trajectory.

As such, the interface brings into effect its own illusory disappearance. This illusory disappearance may be found within the user interface as much as in the study of dynamic natural phenomena. In nature, the interface remained hidden to methods of classification based on the reduction of complex phenomena down to basic constituent elements. Such methods were unprepared to describe or even notice the dynamism and temporality of the interface. It would not be until the latter half of the nineteenth century that the interface was formulated as a condition actively at work in physical processes such as phase transitions, fluid dynamics, and thermodynamics. Likewise, the illusory disappearance of the interface is an essential aspect of the operation of a user interface, inasmuch as an operator internalizes the user interface in the course of working through it, so as to subjectively experience that which is opened up by the interface in a seemingly direct and unmediated way.

At the same time, the interface remains strongly related to the surface, not only in that both entail processes of facing, but also in that each may break down, combine, or be reconfigured into the other. The coming into relation of two or more surfaces may also constitute the production of an interface, while the holding constant of one constituent element in the relation that produces the interface may make of the interface the surface of another constituent element. In this latter case, where the interface becomes the surface of an entity, the entity is still held to the interface as something it must work through in order to possess a face or to be able to act. It is in this way that the interface may in turn be read as a surface of each of the entities it has brought into relation, and so may become a means for the analysis (or close reading, or testing) of these entities. Available to such analyses are the lines of inquiry that follow or recapitulate the penetrations into or colonizations of those entities that have already been accomplished with the coming into being of the interface, and so with the binding together of those entities into relation. In this way, while the interface is not in itself a surface, it may be a producer of surfaces.

Toward a theory of the interface

A theory of the interface is a theory of culture. If culture is an enacted reconciliation of human beings with the social, biological, material, technological, and other realms, the interface describes a cultural moment as much as it does a specific relationship between human user and technological artifact. To use an interface is to participate in culture; it both performs and presupposes an acculturation. Culture here is not given from without but rather produced within moments of encounter. Its elements are likewise not fixed but rather constructed and conditioned according to the relations of which they are part. Along these lines, the interface is more than a theory of interactivity, especially if interaction is viewed as a mediated interplay between stable and self-sovereign entities (e.g., human and machine, designer and artifact, user and control system). This would be to instrumentalize the interface as a fully realizable technology or a soluble problem with respect to a design methodology, within which boundary lines between human and machine are already assumed and given as inviolable.[2] As useful as such compartmentalizations may be in the design of technology, they are broken down in its use and propagation. The issue of how elements produce interaction is inverted in cultural situations to encompass how interaction produces its elements, whether human or machine. The interface occupies this moment of breaking down or inversion, in which elements are drawn into relation and thereby transformed. To questions of interaction it brings questions of agents and agency.[3]

A theory of the interface seeks out culture within the threshold or boundary condition and what may be enacted and entailed upon it. Here, the interface is both imminent and prefigured, just as it at once separates and augments. It describes a zone of contestation that extends from the relation that holds between a user and apparatus to notions of control and power. It brings to bear the questions of what is given to the interface or captured by it, and what in turn is granted by the interface as enhancements or augmentations. As a threshold condition that extends into and incorporates the environments that it bounds, the interface demands of the entities or states that enter into relation with it a surrendering of claims of self-sovereignty and of identities distinct from the threshold, as it is the threshold that becomes the standard by which these are defined and the source from which their agency is derived.

The interface produces a supplementation and augmentation of agency; this is also to say that the interface comes into being as it is actively worked through by its user. At the same time, by imposing itself as a condition necessary for the expression of human agency, the interface comes to define human agency. This defining is also a kind of subject formation or subjectification, in which human agency is brought into compatibility with the interface. While the functioning of a machine interface is certainly defined in part by the machine with which it interfaces, its essential nature lies in being a threshold condition. What most define the interface are the processes by which it draws together two or more otherwise incompatible entities into a compatibility, within which they become available to one another to the extent allowable within the operation of the interface, and from this compatibility produces an overall governance or control.

The interface delimits a specific cultural space, within which a specific set of relations may occur. It presents itself in space and in time, and its operations may be characterized in spatial and temporal terms. Its effects are registered not only in the opening up for access or experience of otherwise unavailable spatialities and temporalities, but also in how space and time are understood and treated within culture. Along these lines, the interface may in part be viewed as a spatial and temporal actualization of those processes of subjectification that characterize the relation of human beings to technology. Here, subjectification occurs in two ways, and involves two distinct notions of space and time. First, in what may be called the *pure* form of subjectification, the operator in held subject to the interface in the course of working through the interface. In its pure aspect, the interface operates as an interior within which the operator is essentially confined during the processes of augmentation. This confinement produces what could be called a *fragmented subjectivity*, in which specific, partial aspects of the operator are identified, enhanced, and entrained within the interface. Second, in what may be called the *active* form of subjectification, the operator acts through the interface, performing what could be called an *augmented subjectivity*. In its active aspect, the interface operates only as a facing toward an exterior; what existed of the interface as an interior condition, in the process of augmentation, seemingly disappears in the performance of that augmentation. For an augmented subjectivity the pure form of subjectification becomes subliminal, a kind of subconscious. That is, the experience of the interface as a form of confinement, an experience only registered by the operator as a not-yet-augmented subject, is subliminal to its experience as a form of enabling by the operator as an augmented subject, for whom the experience of augmentation is incorporated within the hybrid form of an augmented self.

In spatiotemporal terms, the space and time delimited by the pure operation of the interface is folded into or hidden within the space and time of its active operation. In its pure operation, the interface delimits the space that must be traversed for augmentation to occur. It arranges and relates the elements within that space and regulates the timing of its traversal, and so defines the process of augmentation as an event performed in space and time. In its active operation, the interface opens up onto an exterior spatiotemporality that is only accessible to the augmented subject. While the enhanced or altered space and time opened up by the active operation of the interface is seemingly experienced in an unmediated form by the augmented subjectivity, the space and time of its pure operation, which is the spatiotemporality of mediation, still exists within that experience as a kind of subliminal delay. While distinct from one another, each of these notions of space and time, the pure and the active, is bound to the other and could be said to condition the other. The pure operation of the interface is what allows for its active operation, and what defines the territory of enhancement available to be experienced by the augmented subjectivity; the active operation of the interface in turn conditions the evolution of those techniques and processes that define its pure operation as means are conditioned to ends. Following this interrelation, both notions of space and time, the pure and the active, may be found to be in play within the processes of subjectification that occur with the interface. Both are necessary for a description of the interface whether in technical or cultural terms.

Janus and Jupiter

As much as the interface is a problem of agency, it is also a problem of control and power. An illustration of the particular relation between the interface and the wielding of power may be found in Roman mythology, in the controversial figure of Janus and in the relation of Janus to Jupiter (or Jove), the best and greatest (*optimus maximus*) of the gods of the Roman pantheon. Depicted with two or sometimes four faces set in opposing directions, and often with a key in his hand, Janus was the *numen* or spirit of gates, doors, and thresholds. Janus was the god of all beginnings (*deus omnium initiorum*) and of all endings, whose portrayal with two faces represented his ability to look at the same time at both entrances and exits, and into the past as well as the future, whose control over these passages was symbolized by his holding of a key, and who was chosen as the namesake of the first month of the year in the Roman calendar.

In Ovid's *Fasti*, a work composed to illustrate the revisions of the Roman calendar by Julius Caesar, Janus describes his formation as part of a creation myth in the transition from chaos to order: at the time the four elements of air, fire, water, and earth were "huddled all in one," he was "a mere ball, a shapeless lump"; with the "discord" of these elements and their separation into different realms, Janus "assumed the face and members of a god."[4] This primordial transition is retained as an aspect of Janus; so Ovid writes, "the Ancients called me Chaos" (9), and explains Janus's uncanny portrayal with two faces as reflecting this position on the threshold: "Even now, small index of my erst chaotic state, my front and back look just the same." In this sense, in the Roman imaginary through the figure of Janus as a gatekeeper, every threshold opens upon the possibility of chaos as well as of order. "Every door has two fronts," writes Ovid, "this way and that, whereof one faces the people and the other the house-god; and just as your porter, seated at the threshold of the house-door, sees who goes in and who goes out, so I, the porter of the heavenly court, behold at once both East and West" (11–13). At issue here is the question of what kind of power exists in the transition across a threshold; "All things are closed and opened by my hand" is the claim given to Janus, including the exertion of sovereign power, as "my office regulates the comings and goings of Jupiter himself" (11). Janus is honored first among the gods, as "it is that through me, who guard the thresholds, you may have access to whatever gods you please" (15).

In *Archaic Roman Religions*, philologist Georges Dumézil wrote of Janus: "Spatially, he stands on the thresholds of houses, at the doors, *ianitor*, presiding over the two beginnings symbolized by entrance and departure, and over the two which are created by the opening and closing of the door."⁵ The attributes of Janus recall those of the interface, whose visage also watches at the same time in two directions, toward the human sensorium and toward the machine. Like the interface, Janus's realm of control is over the passage between two distinct states, rather than the evolution or development out of a single state. The dominion of Janus extends to the threshold between war and peace, as was symbolized by the *Janus Geminus*, a small shrine called by Virgil the *belli portae* (doors of war), whose state of being opened or closed represented whether Rome was at war or at peace. In reference to "nonreconcilable texts" on the location, number, nature, and even meaning of *Janus Geminus*, Dumézil found an apt comparison for these doors in the Greek myth of Pandora: "According to some, Pandora's jar contained health, according to others, various diseases and ailments afflicting mankind; but when it was opened the result was the same: either health was lost, or the diseases were put into circulation. In the same way, the temple of Janus, when it is closed, keeps or restrains either precious peace or fearful war."⁶ As seen in Dora and Erwin Panofsky's treatment of the many competing versions of Pandora, of whom "no myth is more familiar" though "none perhaps has been so completely misunderstood," both the persistence and malleability of a myth follow its continued usefulness within discourse.⁷ The persistence and malleability of Janus demonstrates the power and uncertainty residing within the threshold.

The mythological attribution of power to the threshold is at the same time expected and difficult to locate and define. Even in Roman times Janus was controversial both in nature and origin: "As to the true nature of Janus, the ancients themselves were puzzled," wrote mythographer James George Frazer.[8] Among the points of contention were Janus's place at the start of the invocation of the gods, preceding the invocation of Jupiter, his apparent uniqueness (there is no counterpart to Janus in the Greek pantheon, for example), and his clouded origins. Frazer's puzzlement with Janus—addressed in the context of his *Lectures on the Early History of the Kingship* and so concerned with the mythological origins of sovereignty—comes down to a question of paternity, since "Janus, like Jove, was regularly invoked, and commonly spoken of, under the title of Father."[9] The attribution of the honorific *Pater* to Janus, and all of the assumptions of sovereign authority that follow this attribution, were for Frazer incommensurable with the theory he found among "some modern scholars, that Janus was originally nothing but the god of doors." Frazer continues: "That a deity of his dignity and importance, whom the Romans revered as a god of gods and the father of his people, should have started life as a humble, though doubtless respectable, doorkeeper, appears to me, I confess, very unlikely. So lofty an end hardly consorts with so lowly a beginning."[10]

To resolve this dilemma, which is one of establishing the origins of power, Frazer approvingly cites the theory put forth by the classicist Arthur Bernard Cook, that "Janus was only another form of Jupiter."[11] Cook, similarly troubled by the fact that "Janus alone took precedence to Jupiter in the divine hierarchy," proposes that "Janus was the name under which Jupiter was worshipped by the Aborigines of Rome ... and that when these Aborigines were conquered by the incoming Italians, their ancient deity Janus and his consort Jana were retained side by side with the Italian Jupiter and Juno."[12] Here the problem of Janus's uncertain claim to power is reconciled by his syncretistic adoption alongside Jupiter, the *Pater* and *Optimus Maximus*; for how could the paternity of Roman power, especially as the Roman religion was the religion of the state, be held by the *numen* of doors and gateways, merely by virtue of his presence upon the threshold?

While Dumézil rejects Frazer's theory of the equivalence of Janus to Jupiter, he remains troubled by the problem of Janus's paternity and power, as "nothing but the god of doors." He proposes instead the sub-servience of Janus to Jupiter, with the control over thresholds in space and time represented by Janus falling under the aegis of the sovereign dignity of Jupiter. Here he cites Augustine in *City of God*, where this defi-nition is attributed to Varro: "Janus precedes Jupiter in recitation, as beginnings (*prima*) are in Janus' power, but summits (*summa*) are in Jupiter's."[13] The power accorded to Janus need not be based on history or conquest, or on the merging of two gods; rather, "whatever attribution of the god we consider derives immediately from his patronage of the *prima* or, as Saint Augustine says elsewhere, from his control over all beginnings (*omnium initiorum potestatem*)."[14] This distinguishing of Janus's power from that of Jupiter is also a subordination; again citing Augustine's quotation of Varro, Dumézil provides an ordering of divine power: "Jupiter is thus deservedly the *rex*, because the *prima* are out-ranked by the *summa*, the only advantage of the former being in terms of time, while the latter are first in terms of *dignitas*."[15] For Dumézil, Janus's position at the beginning of the invocation of the Roman pantheon is not a reflection of preeminence but rather of his position as the thresh-old to that invocation: "If Vesta comes last, *extremis*, in offerings and prayers, Janus is first, as we are told by Cicero. Thus the two of them form the most general liturgical framework."[16] The Janus of the begin-ning is here viewed as one half of the enframing of the Roman pantheon, the other half being Vesta, goddess of the family, hearth, and home; together the deity of crossing the threshold and the deity of dwelling enframe the dignity and the fulfillment of Roman power as represented in Jupiter.

What is at stake here is the distinction and relative ordering of two broad conceptions of the origins of power, each of which produces a particular kind of agency or control. In the first and most commonly understood conception, the source of power, the agency that holds and exerts power, is enabled by that power; it wields power and is set apart by its wielding of power. It is consequently able to exert that power on a substance or entity that it holds external to itself as subject to its power. This subject is subsequently defined, directed, transformed, normalized, and confined by the exertion of power. The agency wielding this power exerts its will as though from above; even in cases where the exertion of power is internalized within its subject, as with disciplinary power, the ultimate source of the exertion of power lies external to its subject. This form of power, in seeking to capture and impose the narrow bounds of an outside order on its subject, removing it from its previous context, is on its own essentially reductive. It may be characterized as the power to define and impose order, as in the power to apply law, and is symbolized in the figure of the sovereign, and in the dignity conferred by the state to the wielding of that power.

In the second conception, power is exerted in the drawing together of once incompatible states into compatibility. Here the agency of power works alongside and within its subject, in part enabling it by rendering the properties of another distinct state of being available for its use, and in part confining it within the processes of its operation. The possible means of operation available to it include translations, transpositions, hybridizations, and phase shifts. It may be characterized by its occupation of a liminal condition, and while it brings about a kind of order, stability, or balance, it is also continuously open to the possibilities of disorder, instability, imbalance, chaos, and turbulence. It could be described as expansive and productive, insofar as it continually seeks out new combinations and hybridizations. It may be symbolized by the moment of the passing over of the threshold, a moment that is now instantiated in the operations of the interface. Here the notion of agency is treated as a central problem; instead of being granted through sovereignty or hierarchy, agency upon the threshold is rather only constituted at the moment of its operation. Where sovereignty refers back to itself in its exercise of power, and distinguishes itself in its wielding of power even when it has internalized itself within its subject, agency upon the threshold continually effaces itself and covers its tracks, making it consequently difficult to assign with any precision the location within which a given decision is made, or to whom or what the authorship of power should be attributed. This second conception of power may also be termed control.

Control and power

While theory of the interface addresses issues of power and causation, it is concerned first with control. Control is not only the final product of the interface, but also the means by which the interface is internally organized and by which it functions. Control describes what takes place within the interface as much as it describes the relationship of the interface to its constituent elements, human or machine. The power exerted by the interface, as a condition that faces or confronts, is directed only toward a bringing into compatibility so that lines of communication or transaction may be established. While this bringing into compatibility may involve an exertion of force or even a coercion, and while it may render a given element or situation available to power relations beyond an establishment of compatibility, it is not yet sovereign power or power properly understood, as it does not yet seek mastery or an attribution of sovereignty. Upon the interface, power is first and foremost local and situational; it is directed simply to establishing and maintaining the conditions of its own existence, conditions by which control becomes possible.

While the interface may be co-opted as a conduit of sovereign power or of the political, this exertion of power is secondary to the interface, which remains primarily concerned with its own internal coherence. In this way the interface presents a complication to that which would seek to exert power through it. Where sovereignty would seek to exert power directly, the passage of power across the threshold alters and supplements it in the course of its transmission. The use of the interface as a conduit for the exertion of power must also contend with the power relations localized upon the interface as a disputed site, as the interface not only enables power but also confines it. In this way there is no political power exerted through the interface that is not also some form of admixture or hybrid, whether material, technological, or human. While a theory of the interface might address power and sovereignty in exploring how the interface occupies the political realm and the ways in which political power is expressed through the interface, it would also necessarily bring up the question of nonpolitical exertions of power and the intersection of these with the political.

Control upon the interface involves a double moment, where power at once confines and enables. Likewise, control involves a doubling of the causal process that is controlled. The notion of control as a supplementation of causality may also be found in the etymology of the word itself: the *contre-rolle* was a duplicate account document created as a verification of an original document and a check against tampering. While created through a duplication, this *contre-rolle* is methodologically more than a doubling. If the act of reading through an original document may be viewed as a kind of trajectory, the *contre-rolle* is a countertrajectory that both runs parallel to its original and intersects it, a text whose meaning only appears when read against another reading. What existed in a first reading is now not only doubly read, but has brought into its reading a supplemental activity of checking or testing, as with, for example, a rapid movement of the eyes back and forth laterally between two parallel though potentially dissimilar trajectories. While the second trajectory of the *contre-rolle* exists by virtue of its having been modeled on the first, it is not for this reason inferior to the first, as a copy may be considered inferior to its original; nor does it, in this initial stage at least, exist to supplant the first. It is rather of an essentially separate type, bringing with it a fully different range of behaviors. The *contre-rolle* exists for reading against and not in order to be read in itself; it does not carry content or produce meaning in itself, but rather exists in a tracking and scanning of the original carrier of content, following the path of its trajectory and testing it at critical points identified along that trajectory.

More than a doubling of the original trajectory of reading, the *contre-rolle* in its essential function adheres as a supplement to the trajectory in order to enable its testing, with the aim of identifying difference. In this sense the *contre-rolle* relies on a kind of parasitic symbiosis to carry out its function. Its duplication of the original text operates at first as kind of mimicry toward this end, just as a model in science is not in itself the natural process that it models, but is rather designed to perform or mimic some aspect of that natural process; the natural process is then said to be understood if the performance of the model tracks it closely enough according to the parameters being used. The identification of difference is important to the notions of understanding and of control, and it is often considered that something is understood or known by the degree it may be controlled. Just as the checking of an original document is allowed by the *contre-rolle*, which tracks it at each critical point, so control in the context of experimental science is allowed by a sufficient hypothesis. Thus one midcentury psychologist would write of the control experiment: "A discoverable fact is a reference or relation, and a discovered datum has significance only as it is related to a frame of reference."[17]

The interface and the apparatus

As a site and means of subjectification that ranges from a general theoretical formation to specific sociotechnical manifestations, the interface bears relation to the apparatus or *dispositif* described by Michel Foucault. A central figure in Foucault's work,[18] the term *apparatus* denotes the concrete means by which relations of power coincide with relations of knowledge to determine what may be seen and what may be spoken. An apparatus comprises both discursive and nondiscursive elements and the connections that hold between them; it is "a thoroughly heterogeneous ensemble consisting of discourses, institutions, architectural forms, regulatory decisions, laws, administrative measures, scientific statements, philosophical, moral, and philanthropic propositions—in short, the said as much as the unsaid." It comes into being in a particular historical moment as a response to a "strategic objective" or "urgent need," where it may be characterized through both a "functional overdetermination" that forces a rebalancing of its interconnected elements, and a "strategic elaboration" that results in effects unanticipated in its original formation.[19] As a theory of power the apparatus represents a cut across society, revealing power relations in a network of heterogeneous and concrete elements whose product is the subject.

While similar to the apparatus as a means of subjectification and method of analysis, the interface neither fully corresponds to the apparatus nor can fully be treated as a subproblem within a theory of power. As for similarities, the interface, like the apparatus, is best understood as having emerged out of a dispersed and heterogeneous set of conditions, developments, and aspirations, rather than from a linear technological progression or out of the interiority of an isolatable idea. It likewise bears within itself an accumulation of techniques, technological and scientific developments, and political, social, economic, and cultural inputs and effects. Both are forms of technological development that hold claim over knowledge and discourse, so raising the difficulty of describing that which already determines the terms of its own description and the channels by which it may be thought. The study of either may be viewed as an archaeology of the edifice on which one already stands, or a genealogy within which one is already implicated. It is in this implication, and as historically situated and presently evolving conditions, that no complete description of either the interface or the apparatus is possible. As with Foucault's earlier notion of positivity, from which *dispositif* would follow, both interface and apparatus may only be treated in part, through a piecing together of fragments that nonetheless addresses a crucial behavior, tendency, or trajectory.[20]

The interface may be distinguished from the apparatus as control may be distinguished from power. This distinction may be found in how each condition treats the attribution of agency and the production of a subject, and in the form of relation each describes. Where the apparatus is a means of tracing the aims and relations of power across society from out of the complications, overdeterminations, and elaborations that arise as power is concretely manifested, the interface describes a complication and entanglement of power: the problem of agency upon the threshold, as a site of a relation and a means of control. Here, the interface does not point toward an attribution of power or sovereignty. Rather, the interface elaborates a multiplicity and indeterminacy of attribution in which agency cannot be fully separated from the relation in which it is engaged. Likewise, where the apparatus describes the production of a subject, the interface describes the production and interrelation of multiple and overlapping subjectivities. The subjectification of the apparatus is most like the pure subjectification of the interface, such that an entity is confined within a relation as a component part, so as to take on a partial and fragmented existence with respect to that relation. Yet pure subjectification operates on both sides of the interface, and also as part of an active subjectification or augmentation. The subjects of the interface are the entities that face each other across the interface as well as the duality of each subject through the simultaneity of a double subjectification, both pure and active, both in fragmentation and in augmentation.

For Giorgio Agamben, the apparatus explicitly becomes a threshold condition defined by processes of separation and contestation, albeit still in a way that may be distinguished from the interface. Expanding on Foucault's work, Agamben envisions "nothing less than a general and massive partitioning of beings into two large groups and classes: on the one hand, living beings (or substances), and on the other, apparatuses in which living beings are incessantly captured." A threshold here is defined by partition and capture. An apparatus for Agamben may refer to "literally anything that has in some way the capacity to capture, orient, determine, intercept, model, control, or secure the gestures, behaviors, opinions, or discourses of living beings." Between living beings and apparatuses is the subject, "that which results from the relation and, so to speak, from the relentless fight between living beings and apparatuses."[21] This struggle, played out in the making and unmaking of the human subject, and between that which captures and that which evades capture or liberates what has been captured, describes the human relationship both with technology and with the state, so that the technological is coextensive with the political. The apparatus draws together what Agamben identifies in an earlier book as the two main lines of inquiry in Foucault's later work: the *political techniques* by which the state exerts control over the life of individuals, and the *technologies of the self* that bring processes of subjectification to bear on individuals, binding them to both an identity and an external power.[22] Subjectification becomes a form of governance, just as the technological becomes the political; this extends to the "reciprocally indifferent" desubjectification of the surveillance state, in which a docile and inert social body takes on a "spectral form" of subjectivity. Agamben rejects the notion that technologies can be repurposed "to use them in the right way"; this would merely be to disconnect technologies piecemeal from their places in the equations and networks of power.

Instead, he presents a conflict ending either in capture or a "restitution to common use of what has been captured and separated" in the apparatus.[23]

Following the work of jurist and political theorist Carl Schmitt on the conceptual likeness of politics and theology as theories of power,[24] Agamben views the separation and subjectification performed by the apparatus in theological terms, with capture as a kind of making sacred, and evasion of capture as a profanation.[25] The apparatus becomes a vehicle of sovereignty, conceived by Agamben as a power above law to determine the sacred exception, or that which is set outside the law. Schmitt defined political power in its most pure form as the ability to determine whether or not law is to be applied. "Sovereign is he who decides the exception," wrote Schmitt in his well-known 1922 essay on political theology; "for a legal order to make sense, a normal situation must exist, and he is sovereign who definitely decides whether this normal situation exists."[26] Likewise, "the concept of the political" of Schmitt's 1932 essay begins with an exclusion, the designation of an enemy, and is so distinguished from all other concepts (e.g., religious, cultural, economic, legal, scientific): "The specific political distinction to which political actions and motives can be reduced is between friend and enemy … the inherently objective nature and autonomy of the political becomes evident by virtue of its being able to treat, distinguish, and comprehend the friend-enemy antithesis independent of other antitheses."[27]

To describe the subject of the political, Agamben draws on the Roman legal category of *homo sacer* (sacred man), who, in being excluded from the law, may be killed without threat of punishment but not sacrificed.[28] What Agamben calls "sacred life" is that in human life which stands outside of law just as political power has been exerted over it in its exclusion. This is also the "bare life" Walter Benjamin described as being defined by the exertion of state violence.[29] Likewise, Schmitt identified "concrete life" or "real life" as the target of the political, in a targeting that seeks out the exception. For Schmitt, "the exception is more interesting than the rule. The rule proves nothing, the exception proves everything: It confirms not only the rule but also its existence which derives only from the exception. In the exception, the power of real life breaks through the crust of a mechanism that has become torpid by repetition."[30] The double act of identification and exclusion constitutes an originary moment of capture; it recapitulates the origins of the claim of political power over human life.

The state of exception is also the threshold of the law, a threshold guarded by the political. For Schmitt, the concern of the political is with the "borderline case and not a routine."[31] While the state of exception does not describe what it means to be lawful or to exist within the law, or even to set forth any rational basis for law, it demonstrates how political power determines the hold of law over its subjects by describing that moment of subjectification whose object is held in a liminal state with respect to the law, as though either at the moment of being drawn into or of being expelled from the law, while yet being fully captured and available to the power of law. Insofar as capture assumes an increase, adding the captured to the already possessed, the sacred or bare life describes not only the operation of political power but also the threshold of its expansion. In this expansion, law is distinguished from the power that establishes it, so that if law is considered rational, the exertion of power is non- or extrarational. Along these lines, Hannah Arendt identified the "pseudomysticism" that characterizes bureaucracy in becoming a form of governance: "Since the people it dominates never really know why something is happening, and a rational interpretation of the laws does not exist, there remains only one thing that counts, the brutal naked event itself."[32] The target of power is stripped of meaning and significance outside of that which is entailed within the exertion of power.

While the sacred represents an extralegal condition, it is by no means extrapolitical; political power is exerted in its purest form on what is captured within its threshold. While held outside social and legal codes, the status of sacred life is still distinguished from natural (or not yet coded) life in its exposure and availability to power. Processes of sacralization or making sacred bring to law the possibility of an expansion, with natural life or life not yet captured, as a territory to be colonized. Political power scans what lies within the law to supplement its internal regulation, and seeks what lies outside the law but may be drawn into the law. In the struggle between apparatuses and human beings described by Agamben, with subjects as an intermediation, the apparatus delimits the political according to its available means, human life its target of capture, and the subject the zone of contestation, delimited as a sacred condition. Against this, Agamben proposes a restoration of human life and community: "Profanation is the counter-apparatus that restores to common use what sacrifice had separated and divided."[33]

The apparatus describes the exertion of power at two scales at once, as a specific and concrete manifestation, often technological, that is also an expression of the political. These two scales describe a tension, as though in a network that recedes and fluctuates in significance to the individual nodes it holds in connection. Here the growing sophistication of the interface puts ever more pressure on the uncertain relation that holds between the technological and the political. The interface departs from the apparatus in its essential irreducibility to a political relation, which defines only one among the forces and materialities that bring the interface into being. In this sense there is no counterinterface, just as there is in the interface no single subject or process of subjectification that may be opposed in a countervailing restoration. For the interface is more than a partitioning, even if this partitioning is also a capturing within a sacred condition or threshold. It is also an augmentation that is emergent from multiple partitionings. In its simultaneity of relations, the interface describes an entanglement of power, agency, and subjectivity, as muc h as it does of the technological and the political. This entanglement is worked out as a continual contestation and reconciliation that produces a state of augmentation. It is the interface that holds its various forces and entities in relation and defines the form of their engagement. Upon the interface, the political finds itself both in conflict and in admixture with the technological, a process played out upon and within multiple subjects. Here, sacred or bare life describes the making of a component to be held in relation; its sacred status is defined most directly by the augmented state in which it takes part. The interface defines human life inasmuch as it has need of it. Likewise, the interface defines the materialities and machines that it also faces; through augmentation, it accords to these as well the sacred status it does to human life. For the interface is the threshold to the machine as much as it is to what is human.

The interface and the game

Just as the interface defines a threshold condition within which an augmentation occurs, so the game defines a game space within which the event of the game occurs. The game, like the interface, is essentially transformable into other conditions; for example, both the interface and the game may become a test or a simulation. Both involve similar processes of subjectification, such that becoming the user of an interface is like becoming the player of a game. And both the interface and the game come into being through a threshold condition; to be the user of an interface or the player of a game is to inhabit and engage that threshold, and to correspondingly be granted a kind of sacred status with respect to the interface or the game. That games and play possess a sacred aspect is evident in the structural similarities between play and ritual. The notion of games as defining a sacred space is common to two foundational texts in the theorization of games and play: *Homo Ludens*, by the cultural historian Johan Huizinga (1938), and *Man, Play and Games*, by sociologist and cultural theorist Roger Caillois (1958). These books share the notion of the sacred as a condition distinct and separate from everyday life, albeit here the sacred is conferred not as a political designation but by entry into the space of the game. For Huizinga, in a widely cited passage: "All play moves and has its being within a playground marked off before hand either materially or ideally, deliberately or as a matter of course. Just as there is no formal difference between play and ritual, so the 'consecrated spot' cannot be formally distinguished from the playground. The arena, the card-table, the magic circle, the temple, the stage, the screen, the tennis court, the court of justice, etc., are all in form and function playgrounds, i.e., forbidden spots, isolated, hedged round, hallowed, within which special rules obtain. All are temporary worlds within the ordinary world, dedicated to the performance of an act apart."[34]

Against a notion of the sacred as solely an exclusion, confinement, or subjection to power, the sacred in play contains the seeming contradiction between freedom of action and confinement within a ruled game space. For Huizinga, the two main characteristics of play are first "that it is free, is in fact freedom," and second, that "it is a stepping out of 'real' life into a temporary sphere of activity with a disposition all of its own."[35] Caillois, explicitly following Huizinga, also begins his definition of play with its being "free" (that is, "not obligatory," so as to retain "its attractive and joyous quality as a diversion") and "separate" (that is, "circumscribed within limits of space and time, defined and fixed in advance"). What Caillois adds to this initial definition—that play is "uncertain" in its outcome, "unproductive" outside the game, "governed by rules," and "make-believe" with respect to real life—are only qualifiers to defining play as requiring both a voluntary freedom of action and a separation from life and limitation on activity.[36] Within play, freedom and the confinement to rules coexist in a kind of tension, in which each opposes the other and yet requires its opposite in order to exist and have meaning. From this tension between freedom and confinement emerges the sacred status that holds within games of all kinds, and according to which cultures may be described according to games they play. For Caillois, "to a certain degree civilization and its content may be characterized by its games,"[37] while for Huizinga, "civilization is, in its earliest phases, played."[38]

To play a game is to enter into a sacred status. A separation from everyday life is common to all of the games and rituals considered by Huizinga or Caillois (including card and board games, puzzles, children's games and make-believe, theater and spectacle, masks and pantomimes, carnivals, athletic contests, individual and team sports, extreme sports, animal races and fights, gambling and lotteries, divinations, ritual trances, shamanistic rites, and religious ceremonies). Caillois's classification of games into four distinct but combinable types—*mimicry*, or simulation; *agōn*, or competition; *alea*, or chance; and *ilinx*, or vertigo[39]—might also be viewed as a classification of forms and experiences of the sacred. It describes the different relations that may hold both internally in the formation of a game, and in the game's orientation toward the world outside itself. What is counted as sacred pertains to each element and action separated into game space. Here the sacred is again tied to notions of power and control. Yet *homo ludens* is not *homo sacer*. The game space may be viewed as an enclosure within which any possible action is already confined or delimited, designed or encoded, judged or classified. It may also be viewed as a condition emerging out of the game actions themselves in their performance, each act bearing within itself that which identifies it as sacred, and so from its instantiation already in a compatible form with any of the other actions with which it relates or comes into contact in game space.

While a given action may possess meaning within a game through its compliance to a system of rules, it also possesses meaning inherent from its onset as an action that comes into being from within the game. As much as the rules of a game may be formalized as a kind of text providing a total description of all activities within the game space, the game as it unfolds in actual play is defined moment by moment as an accumulation and elaboration of actions. None of these actions need carry with them or refer back to a full listing of formalized rules to nonetheless be fully identifiable and representative as actions within the game. The game is not present in the sum of its rules but rather in the acting out or performance of these rules. A limited and confined field of action may then emerge to be a seemingly limitless field of possibility, within which a semblance of almost complete freedom is possible, just as the rules of chess or go are simple in comparison to their practically infinite possible elaborations in play, or as the level of informational complexity found in the human genome stands as many, many orders of magnitude less than its elaboration in a human nervous system, let alone in the living of a human life.

In this way, play may be a separation from life and a confinement within a system of rules at the same time that it may be free or even defining of freedom. What confines within free action is not the game as an enclosure, but rather the game as a threshold that must be worked through in the formation of every action. While the game action does not contain all the rules and permutations of the game, the game action is nonetheless constantly attended to by the game from the moment of the action's onset to the moment of its completion. In this way the free action bears within itself its own separation and confinement; what is more, its freedom has only been possible through the separation and confinement that enabled it. Otherwise it would have neither been a free action nor possessed meaning relative to the game. In this sense, freedom is not, as Caillois argues, constituted within a game as an action "which is free within the limits set by the rules."[40] Rather, like the calculation of statistical or physical "degrees of freedom," freedom is given as a delimited range of action at the moment it is performed as a free action. Within a game, an action is free insofar as freedom is constituted as a condition of constant attention in the playing of the game.

This attending at the threshold defines the game as a form of subjectification. The distinction between pure and active subjectification is useful in defining freedom within the constraints of the game. Pure subjectification is the entrance into the game as a kind of sacred space. A player is the pure subject of a game. In game play, which reconciles the actions of each player as they relate or contest with one another, the player takes on aspects of the active subject, to be defined by the overall activity of the game. And yet the actions of the player, freely acted within the game space, begin from a state of pure subjectification that constantly attends to the player. Here the player at once inhabits and works through the sacred threshold of the game. The freedom of the player, which is without constraint insofar as the player remains within the game as a pure subject, might then be called a pure freedom. In this way the game delimits pure freedom within a state of exception.

The game does not present itself as a condition "against real life," but rather as a form of action that bears within itself its separation, its sacred status. The impetus for Huizinga, Caillois, and others to define play and games as against real life comes perhaps from a desire to distinguish play as an activity wholly distinct from work. For Huizinga play "is never a task,"[41] while Caillois cites as "decidedly and emphatically in error" Wilhelm Wundt's statements that "play is the child of work" and that "there is no form of play that is not modeled upon some form of serious employment." Likewise, Caillois cites Friedrich Schiller approvingly: "Man only plays when in the full meaning of the word he is a man, and he is only completely a man when he plays."[42] Yet today, with society's ever-increasing reliance on computational media, the division between play and work is harder to maintain. Both play and work are increasingly defined by tasks performed in relation with machines and networks.

The elision of the play-work distinction is much like the growing entanglement of the simulation and the real. This entanglement may be found in the question of when it can be said "in real life" that a pilot is flying an airplane. In an airplane simulator a pilot may perform exactly the same actions as in an actual airplane, and while facing exactly the same arrangement of controls and instrumentation. A distinction made here must look outside the actions immediately performed by the pilot, for example to the performance of the airplane or the trajectory of pilot and airplane through space. Differences in the subjective experience of the pilot between simulated and actual flight are not sufficient for this distinction; rather, this subjectivity is only one element within the enacted relation between pilot and controls, and may or may not be relevant to evaluating the simulation with respect to the real. If, for example, controls in reality and in simulation are designed for the remote direction of an unmanned aircraft, such a distinction becomes even more difficult to maintain.

Simulated flight possesses its own ties to the real. It constitutes a real encounter with flight instrumentation that is registered in the sensorium of the pilot and retained as skills applicable to actual flight. The reality of this encounter might even be regarded as enhanced within the simulator, which, if designed for use as a test, might potentially generate a more detailed and accurate record of the actions of the pilot than would the actual instrumentation of an airplane. This enhancement describes an exploration and elaboration of the pilot as a pure subject. It may take the form of a test or simulation, but it may also take the form of an encounter with machines or materialities. Flight is and has always been a mediated activity; even before the airplane cockpit was identified as a distinct spatial enclosure, the central problem of flight was one of establishing the mediations that would allow for the production of control. In this mediation the perception and agency of the pilot are translated through the interface of flight controls and instrumentation to establish a relation with the physical interface of laminar airflow over an airfoil in order to achieve controlled flight. That is to say, an airplane is not directly flown, but rather at a proximity defined by cockpit instrumentation. It is the distance of this proximity that determines the abstraction of an activity into tasks, and so the extent to which these tasks may be transferable, to be reconstituted in other contexts—from flight to simulated flight to gaming, or from work to play—without losing what was essential in defining that activity or in shaping the subjective experience of that activity.

Following Huizinga, the *magic circle* has become a technical term in the theorization of games. For example, game designers Katie Salen and Eric Zimmerman write that "to play a game means entering into a magic circle, or perhaps creating one as a game begins"; and that, "as a closed circle, the space it circumscribes is enclosed and separate from the real world."[43] Here the sacred is again identified as marking the boundary between the real and the simulated; yet this boundary performs far more than a separation. Gaming theorist Jesper Juul has proposed that the magic circle works operationally in the game, marking when the game has entered the realm of fiction; arguing that "the space of the game is *part of* the world in which it is played, but the space of a fiction is *outside* the world from which it is created," Juul proposes that "a series of fictional world spaces with magic circles inside are created and deleted during the course of the game." Here a magic circle operates between an actual encounter with the rules of a game and the fictional space opened up within a game, such that "a statement about a fictional character in a game is *half-real*, since it may describe both a fictional entity and the actual rules of the game."[44] Within this half-reality, for example, the fictional world of a video game may be reconciled with the hardware and software that produce it, just as the tools of literary criticism may be reconciled with the study of computing technology.

Yet games are more than a halfway point between fictions and machines. Media theorist Alexander R. Galloway has proposed, for example, a structural analysis of video gaming that considers not only diegetic and nondiegetic (that is, narrative and nonnarrative) acts within the game, but also whether these are predominantly directed by the player or the gaming machine.[45] Within this framework, all of the actions that make up the video game as an action-based medium—human or machine, fictional or real—are essentially equivalent, comparable, and transferable. In this sense, a video game is defined through a series of reciprocal actions from the player of the game and the game machine, each of which must pass through a threshold of filtering and translation so as to make it mutually compatible. The distinction between reality and simulation fades in importance relative to the status of the actions themselves.

This follows anthropologist Clifford Geertz's description of culture as an "acted document." Bypassing questions of "whether culture is 'subjective' or 'objective,'" or "patterned conduct or a frame of mind," Geertz was concerned instead with "symbolic action[s]," of which "the thing to ask is what their import is: what it is ... that, in their occurrence and through their agency, is getting said."[46] Likewise, the sacred is performed in action, whether as a separation or as a drawing into compatibility in a game. Here the game exists through a continual engagement and working through of its rules and protocols, which describe the formulation of every action. As the player of a game or the user of an interface, one never fully enters into a magic circle; rather one is held at its threshold in an engagement that is gauged moment by moment in the course of action. For the human subject this engagement may comprise any combination of sensorial, cognitive, and kinesthetic perceptions and actions. The magic circle is less an enclosure than a lens that focuses on and engages those aspects of the player or user that are available to action, taking their measure and transforming them into a form compatible to the rules of the game or the protocols of the interface.

The magic circle also describes a kind of interface. In the video game,[47] the game space is situated on the human-machine interface, which informs every action that may await response. The interface reconciles as game play the actions and conditions that define and engage its threshold—whether of hardware and software, of controls and displays, of diegetic and nondiegetic environments, of human faculties and desires, and so on. As a form of sacralization, the interface again brings forth a multiplicity: both in the separation it defines and polices between human and machine and in the augmentation that is at once human and machine. Just as the player accepts a confinement to rules to gain the pure freedom of game play, so the user of the interface is confined to a fragmented and partial humanity to gain an augmented agency. Here the interface occasions a proliferation of new potential identities. It offers this for the price of a concurrent confinement within a humanness that is only a partial reflection of the machine. What is separated as sacred here is not removed from reality to simulation, or from life world to game world, but is rather produced as an operation on the real. Citing Plato's allegory of the cave, media theorist McKenzie Wark argues that the game does not obscure the real as a kind of shadow play but rather actively conditions it. Thus, "gamer theory begins with a suspension of the assumptions of The Cave, that there is a more real world beyond it," since "the real has become mere detritus without which gamespace cannot exist but which is losing, bit by bit, any form or substance or spirit or history that is not sucked into and transformed by gamespace."[48] The interface produces an equivalence of action and communication, if not of substance. It imposes this equivalence, from the rendering compatible of human and machine acts to the blurring of real processes and modeled simulations. It holds humanness up against this equivalence as if to a mirror. As much as the interface comes into being with a voluntary act, it also binds its objects into a threshold of equivalence, and this binding is in itself an exertion of power.

The interface and the machine

The interface is the zone of relation that comes into being between human beings and machines, devices, processes, networks, and even organizations. It exists in the drawing together of elements into a relation and in eliciting from these elements the properties, behaviors, and actions that constitute a state of augmentation, from which control is made possible.

The separation marked by the interface between human and machine is not given a priori, but rather performed in the production of an augmented state. While the interface produces an equivalence between human and machine, through which their actions become mutually communicative, it also distinguishes between human and machine in what it extracts from each of these so that they may enter into relation. For the machine or organization, a sufficient level of intricacy or complexity of operation is required; while for human beings a capability and will toward agency are required, whether trained or untrained, cognitive or sensory, or conscious or subliminal. In occupying the distinction between human and machine, the interface separates itself from technology, whether described as equipment, tools, machines, utensils, utilities, machine tools and automatic machines, computers, or even display systems. While an interface may function as the threshold to one of these forms of technology and the means through which it is put to use, it remains distinguishable from these technologies just as it is distinguishable from its user. The interface is not only the form and protocol by which communication and action occur between technology and user, but also the obligation for each to respond to the other. In this way the interface draws together human and machine into a single, unified trajectory, one that by World War II would be named the "man-machine system." The combined activity of this system or trajectory may be described through its availability to control. For this reason, control could be said to be the primary product of the interface. This remains the case not only for interfaces between humans and machines, but also for other types, as for example between two phases of matter.

The interface introduces a symmetry and reciprocity into the relationship between human beings and machines. The operation of the interface may be viewed as a balancing of the equation that defines and holds together the human-machine system, which allows the translation of the activities of each into a form compatible with the other. The interface is a kind of reification and elaboration of all of the ways human metaphors and behaviors have been extended to the machine and vice versa. The relation it brings into being extends beyond comparative criteria and attributions of likeness, toward a sharing of behavior. In this way, humanness is increasingly defined with reference to machines, just as the evolution of machines follows a path toward ever more sophisticated relations with human beings and societies.

A brief look at two influential and early definitions of machines—engineer Franz Reuleaux's nineteenth-century definition and Lewis Mumford's distinction between tools and machines in his seminal *Technics and Civilization* (1934)—is instructive here. For Reuleaux, called "the first great morphologist of machines" by Mumford, "a machine is a combination of resistant bodies so arranged that by their means the mechanical forces of nature can be compelled to do work, accompanied by certain determinate motions." Reuleaux's definition follows his theoretical work on kinematics, a field he described as a "science of pure mechanism."[49] For Mumford, however, this definition "leaves out the large class of machines operated by man-power" and downplays the long historical development of the machine as a thing tied to its human use; "the automaton," he writes, "is the last step of a process that began with the use of one part or another of the human body as a tool."[50] Mumford distinguishes between machine and tool according to its human use, especially in the figure of the worker. He rejects a distinction based on complexity, as "using the tool, the human hand and eye perform complicated actions which are the equivalent, in function, of a well developed machine."[51] Instead Mumford proposes a human-centered criterion of "impersonality" in use and "specialization" in function, the first having to do with "the degree of independence in the operation from the skill and motive power of the operator," and the second with the "limited kind of activity" available to the automatic machine.[52] For Mumford it is not only that the tool allows a wider range of free action than does the machine, but also that this freedom of action exists as an expression of humanness. As such, the complexity he associates with the automatic machine is of a different quality than the embodied complexity found in the use of a tool. That is, the elaboration of humanness (or of the biological) represents an expansion of possibility, where the elaboration of machineness represents a channeling and a narrowing of possibility.

41

For Mumford the human relation to technology plays out across culture. A "technological complex" that emphasizes the centralization of productive power, the specialization of machinery, and the routinization of tasks brings with it a correspondingly hierarchical, stratified, top-down, and confining form of politics and society. Mumford links the oppressive social conditions of the eighteenth and nineteenth centuries with the technologies of the industrial revolution as a sociotechnical era he calls "paleotechnics"; the decentralizations that followed early twentieth-century technologies of electricity are termed "neotechnics"; and possible future technologies that work in balance with human life are named "biotechnics." (What Mumford calls biotechnics becomes "biopower" for Foucault, a conquest of life by political power through a diffusion of apparatuses.) It is telling here that the dichotomy between machine and tool breaks down in Mumford's third class of technology, the machine-tool, which combines the precision of machines with the need for skilled, trained, and deeply attentive labor. From his example of the lathe to the later development of computer numerical control, the machine-tool represents one evolutionary strand leading to the human-machine inter-face. The machine-tool demands a continuous communication between human and machine, in the pursuit of a task mutually defined through the skilled attention of the operator and the semiautomatic operation of the machine.

Kinematics for Reuleaux would be a "science of pure mechanism," in which the machine is analyzed according to "the mutual motions of its parts, considered as changes of position."[53] A science that "observes changes of position only," kinematics defines positioning as kind of con-trol in which "moving bodies are prevented, by bodies in contact with them, from making other than the required motions."[54] Here an element is positioned in time and space, and in its range of potential motion, by means of a continuous contact or linkage with other elements. Kinematics as a science of positioning recalls the *dispositif* and positivity, both in ety-mology and in functionality. As Agamben points out, Martin Heidegger likewise treats technology as a kind of positioning in his essay "The Question Concerning Technology" (1954); for the technological apparatus,

Heidegger uses the term *Gestell*, a frame or an enframing, defined as "the gathering together of that setting-upon [*Stellen*] that sets upon man, i.e., challenges him forth, to reveal the real, in the mode of ordering [*Bestellen*], as standing-reserve [*Bestand*]."[55] For Heidegger, standing-reserve is all that which technology has rendered available for use; what the machine reveals is the full extent to which humanness is equivalently positioned as a resource or standing-reserve.

The interface may again be distinguished from the apparatus or enframing in its processes of positioning or setting-upon. Positioning or enframing is mutually performed across the interface, and not toward the end of a general rendering available to technology as much as toward the moment-by-moment production of a state of augmentation. The "elements" positioned upon the interface are human as much as they are technological, since the zone defined by the interface extends into each of these to the degree required for its operation. The contact or linkage between these varied elements, through which positioning occurs, is also a kind of communication, whether of information or force. In this way the positioning or setting-upon that takes place upon the interface is entirely rendered in relational terms; it describes the condition of belonging to a relation and so encompasses all of the capabilities and protocols, the communications and transpositions, needed for that relation. There is perhaps no better term to describe this relational positioning, whether in human or machine, than *intelligence*.

Intelligence here refers to a quality determined and adjudicated within a given interface. It describes the capabilities and range of activity of elements brought into relation by the interface, or marked by the interface as territory for further development. Intelligence may be human, technological, social, or material. It is not a fixed property but rather a condition or behavior that is relative to the operation of a particular interface. The situational presence (or production) of intelligence is exactly what defines the capability of human beings or machines to come into contact through the interface. If the primary or external product of the interface and the state of augmentation it brings into being is control, then a secondary or internal product of the interface, as a means of producing control, is intelligence. The state of augmentation brought into being by the interface is essentially a hybrid condition, one equally capable of incorporating electronic sensors and human sensorium, computer processing and human cognition.

The interface is the threshold through which each of the elements, classes, or behaviors it separates can also encounter the other, and through which each acts as a measure of the other. Intelligence, then, is a quality of both encounter and measure. The human-machine interface is neither fully human nor fully machine; rather, it separates human and machine while defining the terms of their encounter. In this way the interface becomes the means by which a human user may encounter its technological other, not directly or in a pure form distinct from human use, but rather through a mediation that already carries with it the conditions of its human use. As a corollary, the human encounter with technology always involves a mirroring. And what is called human intelligence here is a situation-specific and highly contingent admixture.

Likewise, on the other side of the interface, a technological other may be distinguished from technology as incorporated within the interface. Here the concept of the "*machinic phylum,* or technological lineage"[56] proposed by Gilles Deleuze and Félix Guattari is useful. As a theory of nonorganic evolution, the machinic phylum removes the centrality of human need and invention from technological development, to focus instead on a "flow of matter"; or "a constellation of singularities, prolongable by certain operations, which converge, and make operations converge, on one or several assignable traits of expression"; or a "materiality, natural or artificial, and both simultaneously; it is matter in movement, in flux, in variation, matter as a conveyor of singularities and traits of expression."[57] While in part accessible to human making and use, these flows, traits, and singularities are not subject to mastery, but rather must be traced and found in active matter. They are not the products of heroic invention, as by scientist, designer, or engineer, but rather are found by the artisan whose intuition has been focused and brought into resonance by a particular materiality. For Deleuze and Guattari, "this matter-flow can only be *followed.*"[58]

44

The machinic phylum inverts the notion of technology evolving in response to human need, to posit instead the evolution of the human condition in response to the tendencies and flows of matter. Along these lines, technology theorist Manuel De Landa speculates on how a future "robot historian" might write a machine-centered account of self-awareness, describing "the various technological lineages that give rise to their species" such that "the role of humans would be seen as little more than that of industrious insects pollinating an independent species of machine-flowers."[59] Here what is essentially of the machine is wholly other to what is human. With respect to the interface, however, the notion either of humanness inaccessible to the machine, or of a machinic phylum outside of human effect, serves as only as an operational designation of that which has not yet been made available to the interface. Upon the interface, both human and machine are already subject; and that which is not yet subject only points toward a potential availability. Thus a *machinic intelligence* might entail an availability to human use as much as human intelligence might entail a given sociomaterial context. Like human intelligence, machinic intelligence would be determined according to relations that transpire across an interface. One might also speak of a machinic agency or subjectivity, as counterpart to human agency and subjectivity. A distinction might also be made between *pure* and *active intelligence*. Pure intelligence would refer to the capabilities of human or machine as they have been captured and rendered available for use upon the interface; while active intelligence would refer to the channeling of this capability through the interface toward the production of a state of augmentation. Where the interface delimits pure intelligence through a kind of testing, which may also be a colonization, it delimits active intelligence within a performance of control.

The interface describes a fundamental ambiguity between human and machine; it is both a mirror of multiple facings and a zone of contact. This ambiguity bears on the human relationship with technology. For what is first encountered is not the machinic in any pure form but rather the interface itself. The interface is never an object, even in the case of a breakdown or misuse, but is rather a mediated condition that is both inhabited and worked through. What occurs as contact in this working through is a mediated communication, orientation, and decision making between two intelligences that have already been transformed by virtue of their very availability to the interface. In part, what is encountered in the interface is a reflection, albeit a reflection where qualities brought to the interface are selectively transformed and reflected back as both pure and active intelligence. This is an encounter with one's selfhood, now transformed as subject.

Also encountered is a partial aspect of the machinic or technological other, albeit in the form of an active intelligence already rendered compatible within the interface. This encounter is an introjection of machinic intelligence into human selfhood, as well as a projection of human intelligence onto the machine. Yet each of these represents only the internal processes by which the interface constitutes itself and through which it becomes available to function; the function of the interface is in facing, navigating, or controlling an external condition through a process of augmentation. The difficulty in describing the interface follows its simultaneous development of multiple internal and external conditions. It is at the same time a gathering inward and a looking outward, a confinement within a delimited space and a space that enables a delimitation. It entails the problems of human orientation both to and through technology. As a producer of multiple overlapping subjectifications and desubjectifications, internal and external, essentially directed toward threshold conditions and moments of possible transition in phase or awareness, the interface carries with it a tendency to develop and enhance the transformative in that with which it interfaces. In this way the operation of the interface becomes a kind of game of multiplied identities.

Separation and augmentation

The interface may be located in the boundary that distinguishes the modeling and gaming of reality from the real. It is significant here that Huizinga's definition of play through its opposition to the "real" or the "ordinary"—an opposition maintained by Caillois and most subsequent theorists of gaming—is by the end of the *Homo Ludens* presented as a conundrum. In navigating the distinction between play and real life, Huizinga finally appeals to a transcendent ethics: "Whenever we are seized with vertigo at the ceaseless shuttlings and spinnings in our mind of the thought: What is play? What is serious? we shall find the fixed, unmoving point that logic denies us, once more in the sphere of ethics." Or again: "The human mind can only disengage itself from the magic circle of play by turning toward the ultimate."[60] Whereas play first presents itself as separate from reality, it later becomes a very real form of capture. Capture by play is also a vertigo of thought, in which escape comes only *deus ex machina*. Huizinga identifies separateness from real life as one of three main characteristics of play; the remaining two characteristics, the free or voluntary nature of play and the limitations it imposes in time and space, follow upon this separation. The freedom of play is permitted insofar as play "marks itself off from the course of the natural process," and constitutes "a stepping out of 'real' life into a temporary sphere of activity with a disposition all of its own."[61] Yet this separation is not the final definition or teleology of play, as if play were solely an operation executed upon the real; rather, separation is a productive and organizing moment in the operations of a class of activity that includes play.

The game and simulation, though defined according to a separation from "real life" or "natural process," are not oppositional to what they are separated from. Instead, the separation of the game from life describes a becoming available for incorporation and use within a real process. Just as the positing of a thermodynamic interface distinguishes a system from its environment, and so allows a material process to be harnessed to the production of work, so the freedoms and constraints of play find themselves reconstituted in the task. Where play seeks to distance itself from work, it always finds itself conjoined again, inasmuch as each of these involves a separation.

For Huizinga, the separation of play from life produces an ordering: "It creates order, *is* order. Into an imperfect world and into the confusion of life it brings a temporary, a limited perfection. Play demands order absolute and supreme."[62] Yet as much as the moment of separation may be harnessed back to the real, so the absolute ordering of play may begin

47

to order the real, just as the simulation emerges from the real and feeds back into it. In this regard cultural historian Hillel Schwartz describes "the culture of the copy." With respect to the war game, and "the growing coincidence, as much visceral as technical, of *krieg* with *kriegspiel*," Schwartz argues that "simulation and dissimulation are, in our culture of the redoubled event, congeners."[63] If the simulation and the real now possess the same taxonomy, the root of each may be found in an initial separation, through which transformations from each to the other become possible. Thus gaming draws a line from training to performance. For Schwartz, "Drill entrained: it aligned the bodies of the many and made them into one, set them into motion in unison, and geared them to fire regular volleys at close range. Tabletop wargames, repeated day after day, would be the equivalent of drill for officers.... War was chancy, but the wargame, played and repeated over the same grids, was meant, like drill, to leave little to chance on real battlefields where time was more precious than life."[64] The separation of war game from warfare is also the initiation of a process in which drilling and entraining feeds back into the real.

It is the relation of game to war that Huizinga finds vertiginous. At the end of *Homo Ludens* Huizinga writes that "compared with the sham fighting of manoeuvres and drilling and training, real war is undoubtedly what seriousness is to play"; and then, continuing on the same page, that "it is not war that is serious, but peace. War and everything to do with it remains fast in the daemonic and magical bonds of play." Schmitt's friend-enemy antithesis is, for Huizinga, the ultimate expression of that form of separation in which play can no longer be distinguished from the real. Writing in Leyden in 1938, Huizinga links Schmitt's notion of the political—that "all 'real' relations between nations and States" are based on the separation of friend and enemy—to his characterization of war as "*das Eintreten des Ernstfalls*" or "the serious development of an emergency." Huizinga continues: "The term '*Ernstfall*' avows quite openly that foreign policy has not attained its full degree of seriousness, has not achieved its object or proved its efficiency, until the stage of actual hostilities is reached. The true relation between States is one of war." For

48

Huizinga, Schmitt's political theory is "barbarous and pathetic" and "inhuman" but also "correct," albeit only in reflecting the vantage point "of the aggressor who is not bound by ethical considerations." In opposition to Schmitt, Huizinga proposes an "ethos that transcends the friend-foe relationship."[65]

This disagreement may be framed in terms of free will. Huizinga's response to Schmitt is expressed in terms of the liberal political thought against which Schmitt had mounted his critique. For example, the political philosopher Leo Strauss, corresponding with Schmitt in 1932, characterizes Schmitt's granting of the "primacy of the political over the moral" as producing in effect "a liberalism with an opposite polarity." For Strauss, "being political means being oriented to the 'dire emergency' [*Ernstfall*].... He who affirms the political as such respects all who want to fight; he is just as *tolerant* as the liberals—but with the opposite intention: whereas the liberal respects and tolerates all '*honest*' convictions so long as they acknowledge the legal order, *peace*, as sacrosanct, he who affirms the political as such respects and tolerates all '*serious*' convictions, that is, all decisions oriented to the real possibility of *war*."[66] Schmitt criticizes the "anthropological presuppositions" in arguments of free will; he writes, "The fundamental theological dogma of the evilness of the world and man leads, just as does the distinction of friend and enemy, to a categorization of men and makes impossible the undifferentiated optimism of a universal conception of man." In the "methodological connection between theological and political presuppositions," moral theology is for Schmitt an "interference" that "generally confuses political concepts," and particularly "the pessimistic conception of man" and "concrete possibility of an enemy" that underlies all politics.[67]

Here the problem of free will is expressed in terms of both theology and power. The term *moral* operates in a similar way for both Schmitt and Huizinga; it serves as a kind of proxy for the place and possibility of free will with respect to the political, and by extension, to war and to play. For Huizinga, the moral denotes the possibility of free choice that is the only possible way out of the magic, vertiginous circle that circumscribes both war and play: "It is the *moral* content of an action that makes it serious. When combat has an ethical value it ceases to be play."[68] For Schmitt, the moral obfuscates what is at stake in the political, as there is no moral content, no mobilization of "freedom and justice" that is not also available to be "used to legitimize one's own political ambitions and to disqualify and demoralize the enemy." Instead there are only "concrete human groupings which fight other concrete human groupings in the name of justice, humanity, order, and peace," so that "when being reproached for immorality and cynicism, the spectator of political phenomena can always recognize in such approaches a political weapon used in actual combat."[69]

What is described is a game that can no longer *not* be played, a game that has already in the beginning determined the conditions within which all actions or agency may occur and be rendered as meaningful. This remains the case even when, as for Huizinga, some possibility is held open for an exertion of agency outside of an initial separation of play from reality, or of friend from enemy. If Schmitt rejects the possibility of a political moment that precedes the friend-enemy antithesis—holding that the "the high points of politics are simultaneously the moments in which the enemy is, in concrete clarity, recognized as the enemy"[70]—the resulting critique of liberalism, or of any other view that would institute the moral within the political in anything more than an instrumental role, is preliminary rather than direct: liberalism is criticized to the extent that it obscures, confuses, and misrepresents the rules and relations that define the political. For the separation into friend and enemy is the central game mechanic of the political; from this mechanism all other political relations take on their form and meaning. Insofar as this separation is rendered active by the forces arrayed against it on either side, the separation becomes a kind of political interface, from which may spring the sum total of behaviors that are political, up to and including war, in the adjudication or navigation of those energies and forces in collusion and contestation that are the turbulences and flows of all bodies politic. In this, Schmitt criticizes liberalism as an obstacle, not an enemy; as Strauss points out, "the affirmation of the political as such can only be Schmitt's first word against liberalism," as "the battle only occurs between mortal enemies: with total disdain ... they shove aside the 'neutral' who seeks to mediate, to maneuver between them ... who lingers in the middle, interrupting the sight of the enemy."[71] Schmitt's appeal to the doctrine of original sin is not mobilized toward a description of the enemy in moral terms, since the enemy represents the serious in the political. Rather, original sin appears solely to justify the separation between friend and enemy, irrespective of what is designated by that separation, as if by original sin one were thereby consigned (or condemned) to one's place in the playing out of the political.

The subject of the apparatus, the subject of play and the game, and the subject of the political are each produced by a separation. In each case, the enactment of separation brings with it a problem of free will or agency in the face of power and constraint; in each case, separation is an ordering exerted within and upon life, such that the actions or agencies that it permits are only granted meaning with respect to that ordering, whether with respect to the game or to the political.

The interface reframes the problem of free will and constraint. It does not describe one enframing but multiple enframings, whose meaning or reconciliation is given not with respect to the technological or the political, but rather to the performance of the interface itself. If the apparatus is first political and only then active, the interface is first an activity, a governing of those actions that originate from that which the interface separates. For the interface describes an enabling of actions and agencies as much as it does a constraint; it describes an augmentation as much as it does a separation. Where the problem of the apparatus is one of political power, and its expression through discourses, institutions, laws, devices, and so on, the problem of the interface is localized and specific to its own existence: in the separation and binding together of entities whose distinction, even if prior to their encounter with the interface, is produced again in a refigured form, only to be elided in an augmented form through which a mutual action may be directed.

The interface describes a condition in which states of exception are entailed within a state of augmentation. It draws together the entities that it separates in producing an augmentation, and in so doing reconciles within itself the voluntary and the conscripted, the productive and the constrained. With respect to the interface the subject also has a say, even if in a delimited sense, in its subjectification. A positive contribution of action and agency is required from the subjects of the interface, which cannot come into being without an entrance into relation. Here the subject of an initial separation is also voluntarily the constituent of an augmentation, in relation to which the subject emerges piecemeal, in fits and starts, not as a unified being but as a contingent and partial figure, which only comes into focus or definition in providing what is needed to maintain an overall augmentation, and otherwise remains undefined.

What is governed through the interface is also what comes to light in the production of its subject. Thus the player is brought to light by the game, as the enemy is by politics; in either case what is brought to light is only a partial and provisional being, whose existence is contingent to a process of subjectification. Yet the subjects of the interface are multiple and include a fully formed being with respect to its own creation: that of the augmented subject. As with the apparatus or the game, the subject of the interface encounters it with limited agency, as defined by a separation; and yet this subject also stands in relation to other subjectivities, both across the interface and in the spectral and yet seemingly fulfilled subject that emerges from augmentation. The augmented subject is fulfilled in that it encompasses within itself the full activity of the interface; it is spectral in the always immanent possibility of its disappearance with respect to the fragmented subject of separation. While these forms of relation may be found within the political, they may just as well be found within the technological, the material, the natural, the biological, the human, or more likely, an admixture of interrelations between these domains. In this sense the interface is first concerned with the specific conditions of its own contestations, transmissions, and resolutions, and only then with the outside contexts and domains within which the interface comes into use, and within which its forms of relation are expressed as a communication. While the interface may contain a moment of the political, it also produces this moment in reconciliation with other moments, each of which meets within the subject of the interface.

Mimicry in the game and the interface

Upon the interface, each act begins with a fragmentation. Where use to the user, or play to the player, may appear as a seamless experience, this only belies the discrete and multiscalar operations, alignments, and transpositions by which an illusory seamlessness is produced. The fragmented subject pretends to grasp the completeness of the augmented subject, as produced in the event of game play or the full performance of the interface. In this sense the illusory or mimetic function of the interface relates to both an overall performance and to processes by which users, or players, are produced according to an interplay of subjectifications.

The notion of *mimicry* described by Caillois is relevant here. In his classification of games, mimicry is one of four categories, each of which relates to the interface, albeit in altered form. If *agōn*, or competition, is "like a combat in which equality of chances is created, in order that the adversaries should confront each other under ideal conditions, susceptible to giving precise and incontestable value to the winner's triumph"; then *agōn* upon the interface describes a site of contestation in which opposed pressures are brought into resolution, just as hydrodynamics describes a resolution of opposing fluid bodies in fluid form. If *alea*, or chance, is "based on a decision independent of the player, an outcome over which he has no control, and in which winning is the result of fate rather than triumphing over an adversary"; then *alea* upon the interface describes a mobilization of chance, or better, probability, as in the use of stochastic algorithms in the modeling of a complex event. If *ilinx*, or vertigo, is the "attempt to momentarily destroy the stability of perception and inflict a kind of voluptuous panic upon an otherwise lucid mind"; then vertigo upon the interface is in the drawing into relation, the transposition between separation and augmentation, and the distance between the fragmented and augmented subject, just as it is in the testing of the pilot or the entrance or exit into the simulation. Finally, if *mimicry*, or simulation, describes how "all play presupposes the temporary acceptance, if not of an illusion (indeed, this last word means nothing less than beginning a game: *in-lusio*), then at least of a closed, conventional, and, in certain respects, imaginary universe";[72] then mimicry upon the interface encompasses that whole range of techniques by which the simulation is given to correspond with the real, and by which a modeled process is brought into line with its natural counterpart.

Mimicry is also the process by which the partial and fragmented subject assumes for itself the spectral completeness of the augmented subject as it works through the interface. If for Caillois mimicry is present in all forms of play, the same might be said of the interface as it is experienced, insofar as the interface performs transpositions and equilibrations between entities that it has already delimited as separate and incompatible. Caillois's 1938 essay "Mimicry and Legendary Psychasthenia" is also relevant here. In discussing the morphologies and behaviors of insects, Caillois describes a bringing into relation of disparate elements that is also applicable to the interface: "All these details can be brought together without being joined, without their contributing to some resemblance: it is not the presence of the elements that is perplexing and decisive, it is their *mutual organization*, their *reciprocal topology*."[73] This drawing into relation, which includes "mimetic assimilations of the animate to the inanimate," may also be viewed as a form of subjectification. Caillois describes it as a *"depersonalization by assimilation to space,"* where "the body separates itself from thought, the individual breaks the boundary of his skin and occupies the other side of his senses."[74] Mimicry becomes a method of blurring, obliterating, or pushing through the boundary between the organism (or system) and its environment: "The feeling of personality, considered as the organism's feeling of distinction from its surroundings, of the connection between consciousness and a particular point in space, cannot fail under these conditions to be seriously undermined: one then enters into the psychology of psychasthenia."[75]

Such disturbances upon the threshold, and the ontological confusions of mimicry, are standard operating procedure upon the interface. For the interface only comes into being with a discrepancy or difference that it also elides. It steadfastly guards and maintains that difference while at the same time encompassing it within a mutuality or reciprocity. Here mimicry and the illusory describe a simultaneous erasure and enhancement of difference. It is for this reason that the *agōn* of the interface is a contestation as much as a resolution, that the *alea* of the interface is a generation of randomness as much as a gaming of probabilities, and that the *ilinx* of the interface is an orientation as much as an onset of vertigo. Likewise, the mimicry of the interface is also the reconciliation of two distinct ontological experiences, that of the fragmented subject and that of the augmented subject. For the augmented subject, mimicry is not a response to ontology but rather a production of ontology.

The interface also diverges from Caillois's theory of games in his view of the disciplining of play as a progression "from turbulence to rules." Caillois proposes that games be "placed on a continuum between two opposite poles." If *paidia*, a term he derives from a Greek root for child, is "an almost indivisible principle, common to diversion, turbulence, free improvisation, and carefree gaiety," which "manifests a kind of uncontrolled fantasy," then in *ludus* "this frolicsome and impulsive exuberance is almost entirely absorbed or disciplined by a complementary, and in some respects inverse, tendency to its anarchic and capricious nature." For Caillois, "their continuous opposition arises from the fact that a concerted enterprise, in which various expendable resources are well utilized, has nothing in common with a purely disordered movement for the sake of paroxysm." *Ludus*, "in disciplining the *paidia*," subjects the turbulence of spontaneous, disordered action to rules and calculation, so that "a primary power of improvisation and joy ... is allied to the taste for gratuitous difficulty."[76]

The interface presents a problem in which rules and calculation are also the production of turbulence, and where a freedom of action is at the same time delimited in the formation of that action. Turbulence would first be formulated within fluid dynamics as a product of the fluid interface, which would at once define natural fluid flows and render them available to control. In this way the interface denies the possibility of a disordered freedom preceding calculated enterprise, or an imminence preceding organization. For it has already generated a turbulent state out of a system of rules.

For the fragmented subject the interface is immediately experienced as fragmentation, a testing, an extraction. For the augmented subject the interface is immediately experienced as an ability to exert control. It is the interface that calculates the reconciliation of these experiences. Here again the interface performs a kind temporal shift. The interface exerts a kind of retroactive influence over the site at which it is posited, such that its imposition as a means of control takes on the character of a preexistent, anticipated, and inevitable natural occurrence. Likewise, as a process of subjectification the interface is both the immediacy of an encounter and the prefiguration of that encounter in conjoined processes of separation and augmentation. The resultant blurring of subjective experience is recapitulated in the assumption by the fragmented subject of the encounter of the spectral completeness and means of control granted to the augmented subject. Insofar as the operations of the interface may be found in any complex, dynamic process where component elements are bound together into a behavioral coherence or system, so the potential of control is naturalized within any discussion of systems and systemic behaviors, in whatever context or field of operation such systems may be found.

This is not to say that the interface is context-independent, but rather precisely the opposite. While the overall operation of the interface as a form of relation is common to all of the various contexts of its instantiation, the event of its operation is in each instance defined and directed toward the actual and particular in the context or field in which the interface is situated. The interventions that occur upon the interface are in all cases specific, discrete, localized, and situated. The concept of the interface was invented in the nineteenth century as a method of describing complex and dynamic physical processes at a molecular scale, as opposed to a general statistical description. The interface would be means of providing a close-up and intimate description of that which it would control; it would constitute an event within an event.

While the interface anticipated the central concerns of mid-twentieth-century cybernetics, communication, and control, it is also the case that the interface only comes into being as a specific activity at a specific location. Within its given context, the interface actively produces a situated behavioral coherence or embodied intelligence, whether material or social, ludic or political, or as derived from the human sensorium or the workings of the machine. It is particularly in the interrelation of human and machine in cognition and sensing that the interface has, since the start of the twentieth century, come to the fore in ways both hidden and apparent as a cultural form.

The subject of the interface is first a subject of culture. From upon the interface, culture is defined by events of separation and augmentation, and of contestation and reconciliation, through which the incompatible or categorically distinct are brought into relation. In this way culture may be viewed as that which reconciles the material and the social, the technological and the political. While the interface certainly takes its place in the performance of the political, operating in political contexts and employed toward political ends, it cannot be reduced to the political. Yet it may also provide a vital reading of the political, especially insofar as the political carries with it or draws into itself the technological, or what is more, those potentials within and of human beings and societies that are brought into relation to machines through the interface, which would otherwise be unavailable to the political. Here the subject of the interface emerges not as the endgame of a technopolitical process of discipline and normalization but as a by-product and constituent element in a process of augmentation, and yet at the same time as a testing ground on which incompatible realms—machine and human, material and social, technological and political—contest and conjoin.

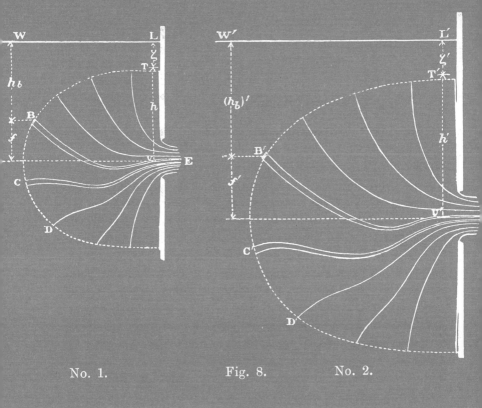

No. 1.   Fig. 8.   No. 2.

2

THE FORMING OF THE INTERFACE

/

The interface as that which defines the fluid

The word *interface* was coined in the nineteenth century by the engineer James Thomson in his influential work on fluid dynamics. It denoted a dynamic boundary condition describing fluidity according to its separation of one distinct fluid body from another. The interface would define and separate areas of unequal energy distribution within a fluid in motion, whether this difference is given in terms of velocity, viscosity, directionality of flow, kinetic form, pressure, density, temperature, or any combination of these. From difference the interface would produce fluidity. As a boundary condition it would be inherently active. While imperceptible in itself, it would be inferable according to its effects. It would be the site of both continuous contestation and the resolution of competing pressures. It would be both internally situated as an existential condition of fluidity and externally directed in the production and harnessing of dynamic form. From its emergence within fluid dynamics, the interface would take on a conceptual affinity with fluidity that extends to all of its subsequent contexts and instantiations.

In notes written in 1869, Thomson describes the formation of the interface as two expansive territories come into contact: "[It is] as if the fluid everywhere possesses an expansive tendency, so that pressure must everywhere be received by the fluid on one side of a dividing surface (or as I call it *interface*) from the fluid, or solid, on the other side, to prevent the fluid from expanding indefinitely, or to balance its expansive force."[1] Here *interface* and *fluid* meet in mutual self-definition. As a technical term *interface* avoids the semantic confusion of "dividing surface," where the use of *surface* immediately brings up the question of how a surface may belong to two fluid bodies at the same time. While *surface* may denote a bounding or enveloping, it does so with a concomitant establishing of an inside and an outside to that bounding. By way of contrast, the bounding denoted by *interface* may be viewed in either of two ways. First, as an internalization of what was previously a boundary facing toward an externality; for example, when what was first given as the external boundary of a thing or condition is internalized as a relation that determines the behavior of a larger flow, assemblage, or system. Second, as an externalization of a facing toward an interiority; for example, when an internal boundary condition that produces the dynamic form or trajectory defining a system becomes either a means of accessing that system from outside it, or a site of influence over a thing outside or over the environment within which it operates.

Further, in opening up an externality within an internal condition, the interface produces, if not a specific form, the potentialities by which a forming may occur. This is just as the phrase *dividing surface* suggests the opening up of a space within the surface itself, within which the potential of division is situated. This forming is a behavior or activity that produces form dynamically in space and time, yielding a static form only if the results of its activities are in some way frozen in time and place. For Thomson the interface would become essential to any description of a fluid or fluid form. His description of the form taken by a flow of water from an orifice focuses on the role of the "bounding interface" in "separating the region of flow with important energy of motion from the region which may be regarded as statical, or as devoid of important energy of motion."[2] Likewise, Thomson relies upon the interface to describe the forming of columnar basalt out of the congealing of cooling lava. This distinctive columnar rock formation, whose regularity could seem the work of a preexisting design, is essentially a random cellular network whose exact form is shaped by factors including the composition of the lava and its rate of cooling. In Thomson's words, the "jointed prismatic structure" of this columnar form follows "a tendency to proceed perpendicularly to successive isothermal interfaces in the cooling mass."[3]

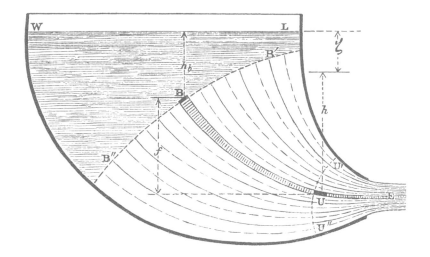

FIGURE 2.1

—

INTERFACE DIAGRAM, 1876. "LET *WL* BE THE
STILL WATER LEVEL, AND LET *B˝BB΄* BE A BOUNDING
INTERFACE SEPARATING THE REGION OF FLOW
WITH IMPORTANT ENERGY OF MOTION FROM
THE REGION WHICH MAY BE REGARDED AS STATICAL,
OR DEVOID OF ANY IMPORTANT ENERGY
OF MOTION."

SOURCE: THOMSON, *COLLECTED PAPERS IN PHYSICS AND
ENGINEERING*, 65.

FIGURE 2.2

—

INTERFACE AS GEOLOGICAL FORM MAKER, 1877.
"EXCELLENT PHOTOGRAPHS SELECTED BY THE
AUTHOR ON A VISIT TO THE CAUSEWAY WHEN HE
WAS SCRUTINIZING THE STONES THEMSELVES."
PHOTOGRAPH BY JAMES THOMSON OF GIANT'S
CAUSEWAY, IRELAND.

SOURCE: THOMSON, *COLLECTED PAPERS IN PHYSICS
AND ENGINEERING*, 429, QUOTE ON 428.

The resulting pattern of fractures in solidified rock renders legible and permanent the operation of isothermal interfaces within cooling and contracting lava. Here the interface inscribes in solid rock the traces of its operation.

Taken together, the interface and the fluid were essential to nineteenth-century conceptions of dynamic form. Dynamic form is less a form than a *forming*, a process active across space and time, and elusive to formal analysis unless captured in some way. Such capture may occur when dynamic form is fixed in time and place as static form; of greater interest is the capture of dynamic form in another important nineteenth-century concept: that of *work*. In this regard the interface and the fluid were instrumental in the development of thermodynamics, following in particular the work of Sadi Carnot and James Joule. In establishing the relation that holds between the generation of heat and the production of mechanical work, thermodynamics would also need to develop concepts of *system* and *environment*. As in the case of fluid dynamics, the interface may not only be used to describe the internal processes by which a system is defined, but also may be found as the boundary that marks the difference between a system and the environment within which it operates. In doing so the interface constitutes the site where a dynamic process of forming may become visible, legible, knowable, measurable, and available for capture in the production of work.

The interface both defines a system and determines the means by which it may be known. It takes its place as the zone across which all activity must occur in order to possess meaning, force, or power. It demarcates the site from which the parameters that define a system may be measured (whether thermodynamically in terms of volume, pressure, or temperature, or otherwise). It is the generative source from which work may be extracted from the system, and the entryway into the system from which influence or control over that system may be exerted. It denotes that part of a system from which change may spring. In defining *system* and *environment*, an interface is drawn into the cacophony of nature, opening up a wild natural process for identification and taming by producing from within it the surface of a system. Here the interface is first imposed as the interiority of the natural process, before being opened up as a surface that demarcates system as inside and environment as outside. In a thermodynamic system, for example, the crossing of an interface is marked by transfer of energy, whether in the form of heat, work, or matter; this remains the case whether such a transfer is assumed to occur within the internal operation of the system, or whether it is extracted as work or dissipated into the environment as heat.

In fluid dynamics or thermodynamics, the interface is a boundary across which dynamic conditions are held in a state of contestation. It elicits a drive to contestation from that with which it interfaces. Thomson identifies this as an "expansive force" inherent to fluid bodies separated by an interface. As the site within a system from which all changes spring, the interface governs change through a seeking of equilibrium. The equilibrium of the interface is a balancing of forces that press against it from all sides, drawn from the entities that it divides. To produce equilibrium, the interface seeks out differential conditions where bodies come into contact. It defines and channels those differences as at once opposing and reconciled within a moment of equilibration. As a boundary and a facing, the interface is in this sense both persistent as an internal form and contingent as a dynamic equilibrium, one that has only just come into being and will at the next moment be dissolved. Its formal persistence exists only in the dynamics of a continual formation, dissolution, and reformation. Within a dynamic form, the interface is not a form so much as a tendency toward a forming, which proceeds through a seeking of difference and its counterpoise in equilibrium.

Along these lines the interface is its own primary product; that is, the interface is first concerned with maintaining its own existence. It does so through the sustained production of momentary states of equilibrium out of disequilibrium. As such, the interface neither belongs to equilibrium nor to disequilibrium, but draws upon each in measure. It owes its persistence within a dynamic form to the maintenance of active contact with the bodies it separates, and it relies upon each of these bodies for the motive force that brings it into being. It exists only through the contestation and communication of these bodies.

The interface does not in itself produce work, though it produces the occasion whereby work may be extracted. In this extraction of work the interface is momentarily transformed into a surface, opening the system so that energies bound up in its interiority are made available outside the system. In this way, work may be viewed as a secondary product of the interface, as may *entropy*. In thermodynamics, entropy, like work, is an extraction of energy from a system; unlike work, this extracted energy is dissipated as heat rather than harnessed as energy. Together work and entropy represent the total energy that may be extracted from a system. The interface is the means by which that energy may be held or dissipated within a system; it is also the means by which energy may be extracted or dissipated from, or interjected into, a system from outside itself through the transforming of interface into surface. Thus both open and closed systems may be described in terms of the interface, which constitutes the site across which all energy transfer occurs.

The interface bears a special relationship with entropy. Not only can entropy, like work, be a secondary product of the interface, but it also can serve, at least negatively, as a measure of the capacity of the interface. In thermodynamics entropy expresses, in terms of loss or potential loss, that energy that the interface would bind into the system. Likewise, in information theory entropy expresses the uncertainty within an information channel that is also its capacity to transmit information. Entropy is that which would be dissipated and unformed if it were not captured by the interface and bound as dynamic form. In a closed thermodynamic system, for example, entropy describes the tendency for conditions that are heterogeneous across an interface to become ever more homogeneous, until a state of total equivalence is reached. This end state is also the complete dissolution of the interface, which only comes into being with a differentiation, and only exists as a moment of equilibrium in the balancing of that differentiation. It is also the "death" of the system, which is now a static, homogenous condition. Thus the second law of thermodynamics, which made entropy into an axiom, seems to lead inevitably toward the universe's end in "heat death." Along these lines, it is the interface, in its maintenance of difference in dynamic form, that constitutes the "life" of a system.

65

In this regard the equilibrium produced by the interface is not a cessation of activity but rather a moment-by-moment balancing of constant contestation. Here, for example, may be cited the "definition of a fluid" proposed by the physicist and mathematician James Clerk Maxwell—a figure whose singular eminence in nineteenth-century science follows the central role he played in both of the major developments in theoretical physics of that century, electromagnetism and thermodynamics. In the fourth and subsequent editions of his seminal *Theory of Heat* (1875 and following), Maxwell defines fluid as that which contains within itself an opposition of forces played out across an interface: "A fluid is a body the contiguous parts of which act on one another with a pressure which is perpendicular to the interface which separates those parts."[4] This definition is identical to those given in previous editions of *Theory of Heat*, with the sole exception that between the third edition (1872) and the fourth Maxwell substituted "interface" for the previously used "surface."[5] Yet neither Maxwell nor Thomson found it necessary to produce a definition of the interface; what the interface was of itself, and the implications of the relations denoted by the interface, would remain tacit within the defining of the dynamic systems (hydrodynamic, thermodynamic, and so on) within which the interface was found to operate. If Maxwell's definition of *fluid* offers what is likely the first scientific definition to include the word *interface*, and the first in which the description of a material state is based upon the concept of the interface, it is perhaps fitting to take as an originary definition of *interface* an inversion of that definition, where the interface is in turn defined with reference to the fluid. Such a definition might read: *An interface is a boundary condition that both separates and holds contiguous as one body those parts whose mutual activity, exerted from each part onto the other, is directed into and channeled across that boundary condition in such a way as to produce a fluidity of behavior.*

While *interface* remained formally undefined, it was defined tacitly with the defining of *fluid*. Thomson's coinage of *interface* in the context of fluid dynamics and Maxwell's subsequent definition of *fluid* according to the presence and activity of an interface, both of which essentially propose the ontological codependence of these two terms, are early indications of the interrelated genealogies that these two concepts share. It is in a more than metaphorical sense that the operation of the interface may be said to be fluidlike, or to resemble the dynamics of a fluid. Likewise, the operation of the interface toward a forming, or toward the production of form, may be said to possess an essential resonance with the forms produced by fluids, whether momentarily instantiated or frozen in place, in the various courses in which they may flow. In tracing the genealogy of the interface in all its various instantiations and contexts, the lines of development followed by the interface tend to follow those courses where what is of concern is that which behaves as or possesses the characteristics of being fluid or fluidlike. The interface is most likely to be found in relation to or within that which is seen as displaying fluidity of form and motion in time and in space: from phase transitions to turbulent flows, from engines to systems, and throughout all those aspects of cultural life where the metaphors of fluidity and flowing may be applied.

In effect, the emergence of the sciences of hydrodynamics and thermodynamics in the nineteenth century brought with it the positing of a boundary condition as requisite to the very possibility of the fluid as a state of being, and to that set of properties and behaviors collected under the concept of fluidity. In its defining of the fluid, the interface provides far more than a criterion for classification; rather, the boundary condition demarcated by the interface is held to describe the moment from which the state of being a fluid and the onset of performances of fluidity are generated. The installation of this boundary condition within the fluid, then, did not constitute a kind of conceptual channeling of the notion of fluidity, as if imposed from without, but rather is given to emerge from within the notion of fluidity as that which defines it, as that from which it springs. In this way, the interface would come to be installed as both the source and site of the genesis of fluidity, and the threshold of that which is fluid.

Turbulence and control

In an early fluid-dynamical description of turbulence, Thomson treats the onset of "transverse currents" as an event situated along the interface of two bodies of water moving at different velocities: "The thin lamina of deadened water will tend by the scour of the quicker going water always moving subject to variations both of velocity and of direction of motion to be driven into irregularly distributed masses; and these, acted on by the quicker moving water scouring past them, will force that water sidewise, and will be entangled with it and will pass away with some transverse motion to commingle with other parts of the current."[6] The interface (or lamina) is the site where the differences between two bodies are registered and then translated into an acting of one body on the other. What results is the production of new states of motion, where the interface is the site of the production of transverse motions, and of the relations of entangling and commingling that hold within a fluid in a turbulent state, as between parts that are separated and yet held together as a single body.

While the interface is the site and condition of dynamic behavior, it does not possess in itself any generative force. Rather, the interface is constituted by and draws upon the force or energy supplied by the bodies that are aligned against it, energy directed outward from these bodies and toward the other bodies with which they seek to contest. This energy is bound up in the interface at the moment of contestation and in the production of dynamic form. Although this binding of energy is impermanent, as is the fleeting form taken by the interface, its persistence across space and time, by which the behavior and activity of the interface gains its coherence, defines the overall state of the fluid or system within which the interface operates. As a momentary, active, and internal binding of energy within a fluid or system, the interface possesses its own distinct identity and rules of behavior, which in turn play out in the fluid or system in defining the dynamic states or behaviors available to it. This momentary binding of energy in a system provides both the identity of the interface and the means of its study; the interface may then be seen to possess its own properties and characteristics, its own processes of formation, dissolution, and propagation, its own rules of transmission and translation, or of opacity and permissiveness. In spatiotemporal terms, the interface expresses the energy that it has bound to itself not only in the dynamic figure that it describes as a kind of surface, but also in the opening up of a space within the system it occupies, from which the spatial relations of the bodies it separates are determined, as well as in the governing of the time of those relations. Here the interface could be said to possess its own spatiotemporality, one that it imposes from within upon the system it occupies. The spatiotemporality of the interface may be viewed as the excess in time and space that must accompany the resolving of forces in contestation, as the interface binds to itself the forces exerted against it and governs their form of transmission.

As a condition that comes into being from within a system, even as it holds together and defines that system, the interface describes a model of control that may be distinguished from those that require externally imposition. In fluid dynamics, for example, the control that the interface exerts over fluidity differs fundamentally from the type of control involved in the channeling of a fluid. In this sense, while the forming produced by an interface may be influenced from outside a system, such influence does not represent direct control but rather an indirect control that must be mediated through the interface, as the activity of forming must still be performed by the interface from within the system. At the same time, the interface opens up opportunities for control from within a system that would otherwise be outside the bounds of control. Here system states and behaviors often considered resistant to control, such as turbulence, are opened up to the possibility of control. Just as the interface is present in all possible system states that may exist in the system that it defines, so that system is also defined by its potential control. In this way, behaviors or states of a system that appear disordered or formless from outside that system may turn out to be well regulated from within the system, and so subject to control. Just as control may be either internally generated or externally imposed, so theories of control must concern themselves with relations that hold between interiorities and exteriorities of control. In the case of fluid dynamics and thermodynamics, it is from attempts to reconcile internally generated and externally imposed models of control that the field of study later known as *control theory* emerged.

The contrast between these two models of control may be found in a thought experiment related by Thomson and proposed by his better-known younger brother and collaborator, the physicist and engineer William Thomson, Lord Kelvin. In this thought experiment Kelvin asked what would occur in a frictionless channel divided down its length by a frictionless barrier with still water on one side and flowing water on the other, at the moment that barrier is removed. For Kelvin, Thomson relates, it was obvious that the resulting state at the moment of the barrier's removal, with a trough of still water in contact with a trough of flowing water, would be "finite" and "essentially unstable." Yet, Thomson argues, this instability due to the absence of an external channeling is not formlessness. Rather, the removal of the physical barrier is also the creation of an internal interface that would thereafter govern each moment of contact between these two bodies. Thomson describes this event as beginning with "the centrifugal motions, or centrifugal actions, which would be introduced on the slightest beginning made of any protuberance or hollow in the originally plane interface between the still water and the current."[7] With respect to determining the onset of turbulence, what stands as externally imposed control is the combination of channel, barrier, and the separation of fluid bodies by that barrier, while what stands as internally generated control is the interface between the two bodies created with the removal of the barrier.

Control exerted through external imposition determines both the conditions and the moment of onset of an otherwise uncontrolled event, just as the onset of turbulence immediately follows the removal of the barrier. In contrast, control that proceeds by internal generation is embodied within the event from the moment of its onset onward. Although it cannot be said to cause the event or to contain it in its entirely, control from internal generation nonetheless inhabits the event, registering its occasion moment by moment in its dynamic form. Thus the onset of turbulence is first marked by the registration of difference in the once smooth faces of interface between the two fluid bodies now drawn together into a single fluid body, and thus the subsequent event is marked by a proliferation of interfaces to describe each transverse motion or eddy constituting that turbulent state.

While the external imposition of control, like the removal of the barrier in the channel, may be viewed as the proximate cause of the event, in that the turbulence would not otherwise have taken place, it does not participate in this event as does the interface in its internal generation of control. Rather, externally imposed control only sets the event in motion as a cause whose proximity never reaches the point of actual contact. Likewise, externally imposed control can never fully be mapped upon internally generated control. This is just as the barrier between the two fluid bodies is by no means equivalent to the fluid interface that separates yet draws them together as a single body, even as that barrier marks the approximate location where the interface is to come into being, and so in that marking essentially serves as a model of the interface. For in dividing the fluid bodies, the barrier relates to them as though they were surfaces, rather than constituting an interface. Yet in holding the place of the interface, the barrier may also model the externally imposed transformation of interface into surface. To extend this thought experiment, a barrier capable of sensing variations of pressure across its length might also be used to extract data that approximate the fluid states of one or the other of the divided bodies. Aspects of the interface that would only be available from within the system are here rendered available, at least in part, as a surface knowledge, to being known from outside that system. Only through such a surfacing is the interface available to some form of close reading.

Even as externally imposed control cannot fully supplant the internally generated control that is embodied within a system, it is the attempt to approach that level of embodiment that defines control theory. What is sought here is to iteratively decrease the proximate distance between the model, barrier, or surface and the event or interface, so as to control a system by shaping and influencing the emergence of interfaces within it. While channeling a fluid body may only require an exertion of power, extracting work from turbulence as a dynamic form requires the continuous engagement of control. It is in this way that the interface becomes a problem of design. Even though a gulf will always exist between the designing or controlling of a system and the actual playing out (or experiencing) of that system from within, it is with this ever-shrinking gulf, the space between model and event, that all fields working through the problem of the interface are concerned. This includes any field that models natural processes, or simulates events, or produces artificial or prosthetic versions of nature, life, or intelligence. As much as the interface governs the shift in system dynamics from internally to externally facing, it also governs the transformation of that system from natural occurrence to artificial process. The problem of the interface, whether in natural, technological, or cultural terms, is of the space that it opens up in such transformations. The aim of control is to occupy that space, the space of interface, and by occupying it, to fill it in or to elide it, or to render it equivocal across the range of its occupation.

It is in this way that the modeling of a thing or process often takes place as a kind of mimicry, where the methods of externally imposed control seek out a more complete resemblance to those that are internally generated. Yet the mimicry of the model, while potentially leading to deeper and more sophisticated methods of control, also carries the epistemological danger of mistaking the model as a form of knowledge with a knowing of the thing or process itself. The more efforts to control a system from an outside vantage point are deemed effective, the more the means of access to that system appear equivalent to the actual events internal to the system. When such apparent equivalence is viewed as more than provisional, one finds the illusory claim that whatever has gained power to externally impose control over a system holds an equivalent control over the internally generated processes that define that system, and so total control of the system itself.

The transformations of interface into surface that accompany the external imposition of control both enhance and complicate this illusory claim of knowing the system itself. Yet control does not seek the reduction or elimination of the illusory. Rather, insofar as turbulent events played out in all their complexity across interfaces in the system remain both essentially unknowable and targets of capture and control, the illusory becomes a necessary and constituent aspect of the knowledge of complex systems. This illusory share of knowing—which, mobilized and sharpened as a technique, is anything but fantastical—exists as a form of play or gaming that increasingly characterizes contemporary knowledge of the world. In this way the distance between model and event may be viewed as a space of play, with the calibration of model to event taking place according to the rules of a game. Likewise, when an interface is an object of design, a space of play exists between the interface as a design problem—including where it is broken down into series of subproblems (or, perhaps, a series of *surfaces*)—and the interface as it operates or is experienced. In this sense the illusory marks all efforts to bridge the gulf between an external vantage point and a system's internal activity.

It is within this bridging, which proceeds through an iterative, back-and-forth, trial-and-error series of ever closer approximations, that the methods of external knowing of a complex system take on a resemblance to the complexity they seek to access. In the same way, a computer model of turbulence possesses in itself a statistical complexity that, while it remains essentially different from the turbulence that it models, still passes far beyond what may be grasped as knowledge by unaided human cognition. Here the statistical model itself becomes another form of interface, as between a computer program and its operator, whose combined product is an attempted knowing of the interface at the heart of the turbulent fluid. Regardless of the statistical sophistication and predictive power such a modeling might have, the epistemological claim it holds over that which it models bears with it an illusory portion, present both in its claim to bridge the model and the event and in its predicting, determining, or positing of causation with respect to that event. The illusory stands as knowledge of the event, in the gleaning of all that transpires in the model-as-interface and in the distillation of calculations that would otherwise have been incalculable to unaided human reason.

This is also to say that the complexities of an event as embodied within an interface, if knowable, are essentially only knowable to a knower that is also product of an augmentation. The problem of the interface thus brings with it an epistemological challenge. If knowledge once sought its categorical foundation in causality, such causal relationships are now only shorthand formalities to processes whose full description is unavailable to an unaided knowing. The interface comes to the fore here as the form (or forming) by which knowledge is produced. As an axiom, it might be said that all forms of the interface share a common and mutual attraction derived from formal similarity. Thus one form of interface becomes the means to know or access another. An example of this is may be found in flight. If a pilot can be said to *know* the interface of turbulent air flow generated by the movement of the airplane's airfoil through the atmosphere, an interface between solid surface and atmosphere whose production and manipulation is the basis for the airplane's controlled flight, such knowledge is only available to the pilot by means of a second interfacing within the cockpit. Here the vicissitudes of laminar and turbulent airflow are translated into an event language accessible to the pilot: a continual stream of sensory and cognitive data ranging from visual cues and instrument readings to kinesthetic and vestibular senses of balance and motion. It is through a productive form of illusion, an illusory knowledge, that a pilot flies an airplane. It is this illusory quantity that bridges the gulf between the operation of a control apparatus and the manipulation of laminar airflow.

The claims of a pilot to possess a "feel" for the aircraft are equivalent to the researcher who claims a feel for a given set of data, or to the user of a software package who develops a feel for what is reflected back through the interface. What is "felt" is the glide of the airfoil through eddying air currents, the smooth statistical sweep of uncountable data points, or the calibrated feedback of algorithms executed at inhuman speeds. In the face of complex and embodied events, knowing is experienced as a kind of feeling—an experienced knowing much like the familiar accounts of the prerationalized knowledge of a craft by an artisan. This reflects not only the gaining of knowledge through active engagement or encounter in a way that renders the full expression of that knowledge impossible outside of the experience of the engagement as an event that is lived through, but also, critically, the fact that the attainment of knowledge is only partially located within a human knower. Or again: as a product that emerges out of the operations of an interface situated between a human operator and a machine, the model represents a form of knowledge the possession of which does not fully belong to human reason, but rather to the augmented condition that comes into being between human and machine.

The exacting of turbulence

Turbulence, a nineteenth-century form of knowledge, has come to typify much of twentieth-century technological and cultural development. While seeming to define a limit as to what may be knowable, a limit beyond which phenomena are viewed as unstable and beyond control, turbulence marks at the same time the potential for the further growth and expansion of knowledge. Turbulence is what appears as the central and most intractable problem of the system; it is the form exhibited by the system at its moment of greatest complexity or at the point of its catastrophic breakdown. At the same time, it marks the path toward a fuller understanding of what actually transpires within a system, of the forces at play and the nature of their relations. Under prior conceptions of rationality, knowledge, and control, the effects of turbulence were designated either as ephemera to be ignored or removed from consideration, or as dangerous system states to be actively avoided or precluded by means of channeling or confining. What would come into being would be new forms of rationality, knowledge, and control capable of tracing more and more closely the movement and relations of these effects, and of finding in both the ephemeral and the dangerous new and productive areas of study.

Relevant here is another thought experiment proposed in 1857 by Kelvin in correspondence with Maxwell. This concerned the problem of the erratic paths taken by falling strips of paper (for both Kelvin and Maxwell, falling strips of paper had been objects of childhood play). Kelvin questions whether this erratic motion could be reproduced within "an incompressible fluid without friction" before suggesting that it could not, since this would make a potentially reversible process out of an inherently unstable phenomenon. In reply, Maxwell disagrees that there is anything inherently irreversible in such complexity: "Now I cannot see why, if you could gather up all the scattered motions in the fluid, and reverse them *accurately*, the strip should not fly up again. All that you need is to catch all the eddies, and reverse them not approximately, but accurately."[8] He continues with an extension of the thought experiment: "If you pour a perfect fluid from any height into a perfectly hard or perfectly elastic basin, its motion will break up into eddies innumerable, forming on the whole one large eddy in the basin depending on the total moments of momenta for the mass." Yet even this "innumerable" proliferation into dissipation of fluid motions is conceptually reversible, Maxwell argues, should the means be found to read each force as it occurs upon each particle, and then to exert these forces upon each particle in reverse. Describing an image that perhaps foreshadows early film's fascination with time reversal, Maxwell proposes: "If after a given time, say one hour, you reverse every motion of every particle, the eddies will all unwind themselves, till at the end of another hour there is a great commotion in the basin, and the water flies up in a fountain to the vessel above. But all this depends on the exact reversal, for the motions are *unstable*, and an approximate reversal would only produce *a new set of eddies multiplying by division*."[9]

In positing the reversibility of turbulence, Maxwell's ambition extends beyond the modeling of turbulence to propose as well its control. Along these lines, control is what bridges the model and the process it models. The moment where control is exerted is the moment where the model and the modeled process are brought into an alignment that, following Maxwell, might be called *exact*. Yet this is an exactness distinct from the exactness of a mathematical solution, or of an exhaustive knowledge of every state of a given system or process, or even of an exact correspondence at all points between a model and a modeled process. The basic principles underlying the mathematical modeling of turbulence, as presently understood, had already been formulated by the mid-nineteenth century, following in large part the work of an older colleague and interlocutor of Maxwell and Kelvin, the mathematician and physicist George Gabriel Stokes. Kelvin was likely referring to Stokes's earlier work on the mathematical description of incompressible fluids when asking whether the behavior of falling strips of paper could be found in such a fluid. The

FIGURE 2.3

—

FLUID DYNAMICS OF FALLING PAPER. DIAGRAMS OF
THE OSCILLATIONS AND ROTATIONS OF THIN STRIPS OF
MATERIAL FALLING IN A FLUID, FROM RECENT EXPERI-
MENTS FOLLOWING JAMES CLERK MAXWELL'S ORIGINAL
WORK ON FALLING STRIPS OF PAPER.

SOURCE: ANDREW BELMONTE, HAGAI EISENBERG, AND ELISHA
MOSES, "FROM FLUTTER TO TUMBLE: INERTIAL DRAG AND
FROUDE SIMILARITY IN FALLING PAPER," *PHYSICAL REVIEW
LETTERS* 81, NO. 2 (JULY 13, 1998): 345–348.

Navier-Stokes equations (so named after engineer and bridge designer Claude-Louis Navier, whose work on fluid dynamics preceded that of Stokes) are widely used today in meteorology and astrophysics as well as in fluid dynamics, from the design of aerodynamic and hydrodynamic surfaces to the computer modeling and animation of fluids.[10]

The exactness Maxwell envisions, however, refers not to the analysis of a turbulent flow by means of a mathematical equation, although this may form part of a means of control, but rather to the potential of acting upon turbulence, of harnessing it and directing it. This in turn requires a kind of presence at the onset of turbulence, a being there at the moment of the event and an acting through that moment. When Maxwell describes such a reversal as being exact and not approximate (that is, viewed as resolvable solely through statistical or other means of approximation), the notion of exactness operates as an absolute only in an illustrative sense with respect to the hypothetical temporal reversal of turbulent flow that he describes. As to the practical problem of describing and controlling a turbulent system, however, exactness is a function of opening an entryway into the system through which it may be acted upon. An exact action upon a turbulent system is at the same time an exacting action, an action that both formulates the conditions for determining what is exact and conforms the system to a state in which it may be known exactly. The exacting of a turbulent system takes place across interfaces, which are the sites within that system through which it may be measured and acted upon.

The notion of the interface, then, does not correspond to an exactness that may be known separately from an act of knowing; rather, the interface finds an exactness only in an exertion of control. Exactness, as a form of meaning, is nothing other than the provisional and contingent bringing into alignment of an externally imposed action with the internally generated processes of a given system. The need to control complex systems required the invention of the interface. The necessity of the interface is precisely the necessity of bridging the external imposition and internal generation of control, and the space of that bridging is precisely the space opened up by the interface. The invention of the interface was the invention of an equivalence, where that which defines an interior condition or interiority is exactly and simultaneously that which allows the opening of that interiority to knowledge and control. An interface theory, then, would pertain to all situations and contexts where control is achieved through an active process that both defines and opens up an interiority.

For Maxwell the reversibility of turbulence would become the problem of control. Even at the time of the 1857 thought experiment, Maxwell viewed the issue of governance, later termed control, as bridging fluid dynamics and thermodynamics. With respect to the reversibility of falling strips of paper or poured pitchers of water, Maxwell could remark: "I do not see why it makes much difference whether these eddies are soon converted into heat, or remain in a fluid state of subdivision."[11] At stake was the inevitable ending of all things in entropy and heat death, as seemingly dictated by the second law of thermodynamics. Control would be Maxwell's bulwark against the triumph of entropy. This would culminate in his best-known thought experiment, first published in 1871 in his seminal *Theory of Heat* and now known as "Maxwell's demon."

For Maxwell the seeming inevitability of rising entropy only follows from the reliance of the second law of thermodynamics on statistical description. His demon, then, would circumvent the statistical description of complex systems. In *Theory of Heat* Maxwell paraphrases the second law as follows: "It is impossible for a system enclosed in an envelope which permits neither change of volume nor passage of heat, and in which both the temperature and pressure are everywhere the same, to produce any inequality of temperature or pressure without the expenditure of work." For Maxwell this need only apply to "our experience of bodies consisting of an immense number of molecules" and the treatment of such bodies according to a "statistical method of calculation." Therefore, the second law "may not be found to be applicable to the more delicate observations and experiments which we may suppose made by one who can perceive and handle the individual molecules which we deal with only in large masses." He calls this latter proposed approach the "strict dynamical method, in which we follow every motion by the calculus."[12]

While acknowledging the practical impossibility of his "dynamical method" to operate at a molecular level within an immensely complex system, Maxwell proposes the hypothetical existence of a particular gatekeeper. He writes: "But if we conceive of a being whose faculties are so sharpened that he can follow every molecule in its course, such a being, whose attributes are still as essentially finite as our own, would be able to do what is at present impossible to us."[13] Maxwell begins with a typical illustration of the second law: an isolated vessel in which half the contents are at a higher temperature than the other half, meaning the molecules in that half possess a higher average velocity. The second law, along with common experience, would dictate a gradual leveling out of difference, toward an end state of equal temperature or homogenous average molecular velocity throughout the vessel. Following the implications of the second law, as Maxwell among many others did in the nineteenth century, yielded a deterministic universe where entropy weighed on all things, and all order was destined to pass into disorder. Maxwell's gatekeeper, "whose attributes are still as essentially finite as our own," would bring a kind of free will or agency to bear against the inevitability of statistical determinism.

Maxwell's gatekeeper, or later, "demon," would take its place at the boundary line separating the hot and cold sections of the vessel: "Now let us suppose that such a vessel is divided into two portions, A and B, by a division in which there is a small hole, and that a being, who can see the individual molecules, opens and closes this hole, so as to allow only the swifter molecules to pass from A to B, and only the slower ones to pass from B to A. He will thus, without expenditure of work, raise the temperature of B and lower that of A, in contradiction to the second law of thermodynamics."[14] Maxwell's demon has been the subject of contention from the time of its positing to today, not only with respect to the second law of thermodynamics, but also more generally to the meaning of statistical descriptions and even to issues of free will and determinism. Many of these debates turn on the agency attributed to Maxwell's demon, "whose attributes are ... as our own." A definition of *demon* in this sense was first published by Kelvin in 1874: "A 'demon,' according to the use of this word by Maxwell, is an intelligent being endowed with free will, and fine enough tactile and perceptive organization to give him the faculty of observing and influencing individual molecules of matter."[15] Maxwell's demon situates control over complex, turbulent processes at the level of the particular, and through this control seems to open a space of freedom in the face of statistical inevitability. It thus takes on a particular resonance to theorizations of flow, where it offers the possibility of deviation and reversal. This may be seen in an exchange between composer and polymath Iannis Xenakis and the philosopher Michel Serres in the context of Xenakis's 1976 thesis defense at the Sorbonne. Xenakis describes himself as "possessed by Maxwell's demon" in the pursuit of "orders which can be outside time" and so reversible; these orders are "not in real time, meaning in the temporal flow, because this flow is never reversible," but rather "in a fictitious time which is based on memory." For Xenakis, "Maxwell's demon can reverse things." Serres then asks: "So, there are reversible structures in music." Xenakis: "They are reversible outside time."[16]

The demon on the threshold

The ancient Greek *daimon*, meaning spirit or divine being, was derived from a Greek root meaning to distribute or divide. That this dividing also involved a discerning is suggested by Socrates' use of *daimonion* to refer to the inner voice that controlled his thoughts and actions. For mythographer H. J. Rose, *daimones* were first "merely a vaguer equivalent of 'gods'" but came to occupy an intermediate position between the earthly and the divine. After Hesiod, Rose writes, the word *daimon* tended "to signify super-human beings of something less than divine rank," whereas after Plato, "their proper abode is neither heaven, which belongs to the gods, nor the earth, which is the home of men and the lower animals, but the air, which lies between heaven and earth. Corresponding to this intermediate dwelling-place is their intermediate nature. They are superior to men, inferior to gods."[17]

By the first century B.C., the Latin *numen*, which had originally referred to movement or motion, took on a religious denotation as a term corresponding to *daimon*.[18] At this time in Greco-Roman theology, both daimon and numen were used to refer to "an inferior class of divine beings, for those who dwell in the sublunar sphere or even on the earth, the nymphs, fauns, pans, etc."[19] The kind of spirit denoted by *daimon* or *numen* (or alternatively by the etymologically related Latin *genius*) has to do in particular with the spiritual identity of a material thing, its proliferation or procreation, and the animation of inanimate things. Rose writes that "even inanimate things, if there is something holy or uncanny about them, may have or even be *numen*."[20] The attribution of *numen*, or of the *daimon*, may be seen as adhering to, or residing within, those things or places whose particular qualities appear to reflect the working of an intelligence that is nonetheless not fully intelligible, or of a will that is in some way comparable but not ultimately reducible to human will. The *numen* or *daimon* may be appealed to, albeit without assurance of success; it stands for that which must be engaged in the hope of controlling situations or processes that otherwise elude understanding. Clarifying the uses of *numen* in the Roman religion, Rose says that "*numen* is needed to produce offspring; *numen* can attach itself to certain inanimate things; the table has its *numen*; the house, parts of the door, were laden with *numen*; the fat of the wolf used to anoint the door-posts was heavily charged with *numen*; sexual relations are full of *numen*; *numen* is soaked into the doll of Compitalia [an annual festival celebrating the household deities of crossroads]; and so on."[21]

The terms *daimon* and *numen* (to which could also be added the Roman *genius*) each represented a way of knowing and a possibility of control, as well as the limits to that knowing and the slippages and blind spots in the exertion of that control. As Greco-Roman religion came under increasing pressure to rationalize its beliefs and rites, the *daimon* or *numen* would come to represent the uncertain relation that holds between the spiritual and material realms. For Plutarch among others in the first century A.D., Rose relates, the *daimon* became useful in explaining those aspects of religious belief and practice found to be irrational or out of keeping with an understanding of the divine as rational, perfect, or transcendent. Here *daimones* would take the place of gods in those cases where "there were not only myths representing some of them as behaving in a way quite inconsistent with any developed ideas of deity, but rites which seemed to aim at propitiating unfriendly powers and inducing them, not to do any good, but simply to refrain from doing harm."[22] As the rationalization of Greco-Roman religion sought to privilege notions of the divine as transcendent to material relations, the *daimon* and the *numen* came to stand for the often messy relations between the divine and the material as handed down in myth. Rose continues: "If a myth, authoritative through its age or its association with venerable rights, was morally unpleasing, it might still be accepted and the believer's conscience rest undisturbed by the simple assumption that it referred to *daimones* and not to gods proper. The former might indeed, being ethically imperfect, fight one another, make love to mortal women, be banished from the society of their kind for their offences, or even die, none of which things is becoming to divine majesty."[23]

The *daimon* or *numen*, then, found its sphere of influence increasingly identified with that which was uncanny or unknowable in a given materiality. At the same time, as the animate within the inanimate, it offered the possibility of an often illicit or ambiguous knowing or control of that materiality. As the gods withdrew from the material world, the *daimon* became increasingly associated with the practice of magic and the telling of oracles. Thus, writes Rose, if a myth of Apollo portrayed him seeking revenge, it was not the action of the god himself, but of a *daimon* that took his name, and "if his oracles ceased, as for a while they showed signs of doing, the reason was entirely to his credit, for he was becoming so exalted that contact with matter was no longer possible to him."[24] Poised between the earthly and the divine, the *daimon* or *numen* opens up the possibility of communication between separate realms at the same time that it threatens miscommunication or deception. As the spirit present in material things, processes, and events, it makes possible otherwise impossible manipulations of the material world. It is in this sense an inverse of the Platonic form: instead of a form of knowledge where particular, material instances point toward eternal and unchanging forms, the knowing of the *daimon* proceeds from an animating form or spirit toward its material instantiation. In early Christian apology, particularly in its dialogue with Greco-Roman religion, "demon" began to connote a malicious, deceitful, and evil spirit, capable of possessing and controlling things and beings and encountered not in abstraction but with a specificity of situation, location, and event.

Thus the *daimon* or *numen* stood as kind of animating spirit of intersections of all kind. It was a mediator between the material and the divine, a spirit inhabiting thresholds and crossroads, and even, in its proliferation and attendance to procreation, a bringer of transformations in state or of the genesis of new forms. At the same time, it would be bound to particular material instantiations, whether things, processes, or events. To return to Maxwell's demon, it is fitting that Kelvin introduces the demon in its place of inhabitation: the interface. The introduction of Maxwell's demon at the interface takes place in what is also perhaps the first published use of *interface* outside of the work of Thomson and Maxwell. In place of Maxwell's vessel, Kelvin's description specifies a perfectly insulated and uniform metal bar at an initial state in which one half of the bar has a different temperature from the other. Against the second law's prediction of an eventual final state of uniform temperature across the bar, Kelvin writes: "This process of diffusion could be perfectly prevented by an army of Maxwell's 'intelligent demons' stationed at the surface, or interface as we may call it with Prof. James Thomson, separating the hot from the cold part of the bar."[25]

Kelvin defines these demons according to their agency. Here they possess the ability to measure the energy and course of individual molecules; the ability to reverse the course of molecules while maintaining the molecules' energy ("every time he strikes a molecule he is to send it away with the same energy as it had immediately before"); the ability to occupy a specified place ("every demon is to keep as nearly as possible to a certain station, making only such excursions from it as the execution of his orders requires"); and critically, the ability to follow a given range of orders—to be programmable. The interface, then, is reconfigured as an occupied zone: "The whole interface between hot and cold is to be divided into small areas, each allotted to a single demon. The duty of each demon is to guard his allotment, turning molecules back or allowing them to pass through to the other side, according to certain definite orders."[26]

While Maxwell did not directly specify his demon as occupying an interface, this occupation is tacit within *Theory of Heat*. For example, in a discussion of how "to obtain a distinct conception of the flow of heat through a solid body," Maxwell proposes a "surface or interface to be described within the body such that at every point of this interface the temperature has a given value." He describes this interface, following Thomson, as an "isothermal interface." Maxwell describes the flow of heat or thermal conductivity through a solid body as a communication determined by the interaction in space of these isothermal interfaces, distributed through the solid body as a geometric series of nested shells whose transmission of heat follows the definition of fluidity in flowing perpendicularly to the interface.[27] Kelvin's description of demons occupying the interface only renders explicit what was implicit for Maxwell. Now occupied by demons active according to their capability and programming, the interface not only circumvents the second law of thermodynamics, but also makes conceptually possible a fully programmable and "intelligent" materiality. Kelvin demonstrates how giving different orders to the demons on the interface in terms of selection and permission could yield different system states. He proposes a program for impermeability, a program that would maintain a temperature differential across the interface while still allowing a normal diffusion of energy, and a program that in a gaseous system would maintain equal temperatures on either side of the interface, though with one side consisting of high energy molecules at a very low density, and the other of low energy molecules at a very high density.[28]

The interface is in this way activated by the presence of demons. It takes on the qualities of limited agency and directed intelligence according to the programming of the demons that inhabit it. While the boundary zone delimited by the interface could certainly be the site of activity without such demons—just as the interface in Kelvin's metal bar would be the site of a statistically describable diffusion of energy, and the interface in Thomson's channel would be the site of the production of turbulence—

85

it is only through the occupation of the interface by demons that such activity becomes subject to control. Here demons perform the bridging through which the internally generated interface is transformed into an externally available interface or surface, and the internal processes and behaviors of a system are opened up to the possibility of control. Kelvin's configuration of demons to produce an impermeable barrier could also be seen as a first step in the transformation of that interface into a surface; should the demons be granted the ability to count and measure the velocities of the molecules whose trajectories they reverse at the interface/surface, the result would be a measurement or testing (of temperature, density, etc.) of the condition inherent to the interface/surface.

The "intelligence" and "free will" attributed by Maxwell and Kelvin to the demon are essentially constrained both by its programmability and by its given range of capability. Yet at the same time, the demon is granted a sort of lifelikeness, a semiautonomous agency that mimics intelligent behavior and the performance of free will. The demon stands as a kind of choice within complexity and statistical inevitability, even if this choice is fully constrained and exists only for the purpose of control. Here, again, the demon is that spirit which resides within a materiality, and which is appealed to in the knowing of what is otherwise unknowable, and the controlling of what is otherwise uncontrollable; and here, again, the demon carries with it an inescapable aspect of uncertainty and the uncanny. For the demon is set to work at its task and the parameters of its behaviors are defined, but the event of the demon's labor is its own.

Just as the concept of entropy spread from thermodynamics to information theory, so Maxwell's demon would become a useful figure in computer programming. The term *daemon*, chosen specifically in reference to Maxwell's demon, was first adopted in computer programming to designate a kind of background program or subroutine. It came into use in 1963 in the course of MIT's highly influential Project MAC, a project begun with an ARPA grant under the direction of J. C. R. Licklider. Project MAC included many influential figures in computer science and achieved breakthroughs in computer networking, operating systems, and artificial intelligence. Today *daemon* is defined as "a memory-resident program that remains in the background, awaiting an event that triggers a pre-defined action"; or as "a program that runs continuously in the background or is activated by a particular event"; or as "a background program ... that performs housekeeping and maintenance functions automatically." Or again: "A daemon sits in the background and is activated only when needed, for example, to correct an error from which another program cannot recover."[29] The daemon, then, resides in a background with respect to an actively attended-to process. Continuously active in its waiting, it seeks out the predefined conditions or event that would trigger its performance of a predefined activity. After being programmed, the daemon does not require active attention in order to function. The daemon is defined by its capability to perform its own attending.

The daemon is an aspect of a computer interface that attends precisely when it is not attended to. The daemon forms part of what could be considered the subconscious or tacitly known experience of an interface by its user, even as that user consciously works through that interface toward another object of attention. Waiting in the background, at the margins of attention, the daemon acts as a supplement to the user's conscious use of the interface. At the same time that it grants to the user the ability to consciously exert control, however, it also represents the extent to which such control relies upon hidden background processes. The daemon represents the portion of uncertainty embedded within controlled certainty; it stands for the anxiety of losing control that accompanies every increase in control. As a symbolic processing that mimics human cognition and agency, even while being highly constrained in its programming and superhuman in its speed and accuracy of calculation, the daemon on the interface is tacitly engaged but not consciously knowable. The space between engagement and knowing is for the user the potential for anxiety and surprise. As one of the first and perhaps the greatest of computer programmers, the mathematician Alan Turing, wrote in 1950: "Machines take me by surprise with great frequency."[30] Even the first computers were capable of producing surprise, of exceeding the expectations of their users. In possessing this capability for surprise, computers demand to be *engaged* in their use as opposed to simply being *known*.

For Turing this involved recognizing the limitations of human cognition as it engages with the raw computational power of computers. Surprise occurs "largely because I do not do sufficient calculation to decide what to expect them to do, or rather because, although I do a calculation, I do it in a hurried, slipshod fashion, taking risks."[31] Here Turing draws attention to the gulf that exists between unaided and augmented calculation. With respect to the human-computer interface, this gulf is the space opened up for the operation and inhabitation of demons, just as the daemon represents an attending to the interface that is distinct from the user's attending to the interface even as it conditions the user's experience. That the daemon, along with the gulf that exists between human experience and the experience of augmentation, may be hidden from view or actively ignored follows the extent to which neither of these appears on the surface of things, but rather only in the internal working through of processes. In Turing's words: "The view that machines cannot give rise to surprises is due, I believe, to a fallacy to which philosophers and mathematicians are particularly subject. The assumption is that as soon as a fact is presented to a mind, all the consequences of that fact spring into mind simultaneously with it.... A natural consequence of doing so is that one assumes there is no virtue in the mere working out of consequences from data and general principles."[32]

The "mere working out of consequences" is what is at issue here. It is only in a working out of and through an interface that control is possible. To return again to Kelvin's treatment of Maxwell's demon, the occupation of the interface by semiautonomous agents allows the possibility of knowledge and control within situations of otherwise unintelligible complexity: "By merely looking on crowds of molecules, and reckoning their energy in the gross, we could not discover that in the very special case we have just considered the progress was towards a succession of states in which the distribution of energy deviates more and more from uniformity up to a certain time."[33] If statistical treatments of complexity tend toward their homogenization, the demon and the interface open up hidden differentiations. Control is from the beginning a resistance against temporal inevitability and homogenization; just as this resistance was first expressed in a positing of the reversibility of time, so all control is essentially temporal in nature. Further, control provides a capability to work within and upon local scales toward the end of influencing a global or system-wide behavior that would otherwise be incomputable, inaccessible, or overwhelming in statistical complexity. Control is a kind of legerdemain, a cheating of time, whereby one form of uniformity—that given as the ultimate outcome of chaos, turbulence, and diffusion as subject to the inexorable gravitational pull of entropy—is replaced by another uniformity, as imposed by control. As much as control finds, exploits, and produces difference, it does so through imposing a uniformity between interior process and exterior access. This uniformity is the means by which the interface comes into existence, and so of any kind of communication across the interface.

Yet this imposition of control carries within itself the seeds of another form of chaos, one that is located in the material particularities and events that must be navigated, processed, or filtered in that imposition. While all control is to some necessary extent localized and particular, this sense of locality and particularity is itself an artifice of mediation; it proceeds as though it were an appeal to a spirit or intelligence—a kind of *genius loci*—that remains to some critical extent foreign to the conscious aims of the exertion of control. Instead, the demon in the interface, as a kind of animation of materiality, partakes of that irreducible materiality as much as it mimics a kind of humanlikeness and submits to human will, so establishing within the moment of control an element of the uncontrollable.

Theories of the vortex

Before the interface was the vortex. Conceptually and technologically, the problem of the vortex gave rise to the positing of the interface. To extend this genealogy, it could be said that the vortex is situated on the threshold of control. Until the invention of sciences of control, the vortex stood at the point where control—or rather, the relatively limited models of control represented by the channeling of water, the navigation of seas, and so on—was an impossibility. If order is viewed in Euclidian terms, as a seeking of straight lines between points, the vortex—the whirlpool, *tourbillon*, turbulence—is the danger of a spiraling and vertiginous disorder. As opposed to the clearly set course of celestial navigation or the linear geometries of the aqueduct, in which the placidity of water was sought after and delimited, the vortex represented a kind of elemental, material rebellion against all ordering and planning. It was the upsurge of confusion, entrapment, and destruction that philosopher Gaston Bachelard characterized as "violent water,"[34] the cataclysm of storms, and in mythology the Charybdis from which Odysseus makes the narrowest of escapes.

Yet if the vortex represented a kind of disorder, especially with respect to human plans and convenience, it was also a compelling order. Turbulence produces remarkable forms, including the vortex. The vortex is not only destructive but also innately and prolifically creative. It is not surprising, then, that turbulence and the vortex should suggest an alternate order, one opposed to an order represented by planning, regulation, confinement, and channeling. The vortex is an ordering produced by a materiality of flux. For Bachelard, images of turbulence, as with all images derived from water and fluidity, stem from a "material imagination" where flux (following Heraclitus) is shown to be "a *concrete* philosophy, a *complete* philosophy," and where "water is truly the transitory element.... A being dedicated to water is a being in flux."[35]

Along these lines, turbulence and fluidity would be essential to distinguishing the opposing organizational regimes proposed by Deleuze and Guattari: the State as opposed to the nomad, or the striated as opposed to the smooth. For Deleuze and Guattari, both State and nomad possess a distinct and characteristic "hydraulic science." Thus: "The State needs to subordinate hydraulic force to conduits, pipes, embankments, which prevent turbulence, which constrain movement to go from one point to another, and space itself to be striated and measured, which makes the fluid depend on the solid, and flow proceed by parallel, laminar layers. The hydraulic model of nomad science, on the other hand, consists in being distributed by turbulence across smooth space, in producing a movement that holds space and simultaneously affects all of its points, instead of being held in by space in a local movement from one specified point to another."[36] Turbulence is the test case here. State science is founded on the prevention of turbulence, and so on the pacification of fluid behavior by means of constraints, measurements, and channeling. Fluidity is held subject to a rationality characterized by the grid and the striation. For nomad science, turbulence is a source of creation and of "distribution"— a term adopted from the work of the naval officer turned philosopher-poet of fluid form Michel Serres. Turbulence produces an immediate spatiality: one which emerges from within fluidity and without resort to imposed striations; which operates according to temporal simultaneity and smooth spatial transformations, not a breaking down into sequence or grid; and which is aligned with a vitalist rather than a mechanistic form of rationality.

This opposition is often posed as one of free action against constraint. Serres invokes the Roman poet Lucretius, who in his epic poem *On the Nature of Things* opposed free will to determinism in atomistic terms. Here the possibility of freedom arises from the *clinamen*, or swerve, of atoms into unexpected trajectories. In a passage cited by Deleuze and Guattari, Serres writes of "that world of physics in which the conduit is essential, and the *clinamen* seems like a freedom because it is a turbulence that rejects forced flow."[37] For Serres the vertiginous and unsettling vortex is also the source of freedom and possibility, insofar as it sets into time and motion that which was constrained in space. "The swerve brings time into existence, it produces it…. It throws us off balance, makes us unsteady. Hence we are on the move."[38] The swerve is a distinctly fluid property; it is aligned with turbulence and the vortex as it is with chaos: "Chaos is open, it gapes wide, it is not a closed system … it is a multiplicity. It is multiple, unexpected. Chaos flows, it flows out, an *Albula*, a white river."[39] It is in this way aligned with life: "The *sea* is the cycle of life. One understands here why Lucretius invokes it in the beginning: fluid originates the distribution from which every living thing first emerges."[40]

While fluidity and the swerve stand as the genesis of movement and life, they are also for Serres essentially resistant to thought: "Look how much trouble we have thinking or seeing it. The whole reason protests—I mean, logically. Our whole classified rationality, all the coding, habits and methods, lead us to speak in externals or negatives: outlaw and nonsense."[41] Thought rejects the open, seeking confinement instead within the closed system: "Philosophers of the contemporary age are philosophers of the reservoir. Of the circulation of things stored in the reservoir."[42] Serres's project instead seeks the open through a thinking of the vortex. Yet what Serres holds up against the stratified order of the closed system is not disorder, but rather a mediation of order and disorder. To the question "What is turbulence?" Serres replies, "It is an intermediary state, and also an aggregate mix"; or again, "Turbulence is an intermittence of void and plenitude, of lawful determinism and undeterminism."[43] Serres finds this intermediation in the very words turbulence and *tourbillon* (vortex), in the play between the Latin *turba* and *turbo*. While both refer to a kind of dynamism, the first is the turmoil of the mob, while the second is the dynamic form of a spinning top or spiraling whirlpool. If *turba* connotes an essential formlessness, *turbo* connotes the seeking of stability and balance in constant movement. Thus Serres writes of "the *turba* of Lucretius, a stormy mass of diverse elements in disorder, given over to shocks, to impacts, to the fray, a chaos given over to jostling, is a crowd, it is a mob. The physical chaos of circumstances, where the primal *turbo* spirals itself along, is, if I may be so bold, isomorphic with the raging crowd."[44]

In this sense, what stands as the open for Serres, and as resistance against the imposed and reductive order of the channel and the reservoir, is not an absence of rules, forms, or order. Rather, it is a moment of negotiation between chaos and order, between dissipation and work, between the formless and the formed. Each of these is held in tension with the other, and either one may emerge from a moment of negotiation to dominate for a time the shaping of an environment, whether toward turmoil or a dynamic stability. The free and open emerges from a condition of negotiation between the constrained and unconstrained, and it is this condition that is at the heart of creation and life. Following Serres, the problem of the vortex is of its position poised as a turbulence that contains within itself both *turba* and *turbo*, the vertiginous mob and the self-sustaining balance of the spinning top. For the moment of negotiated equipoise represented by the vortex is also the site and means of control. The invention of the interface was precisely a theorization of that moment of equipoise within the vortex, as the means of access by which its passage from chaotic dissipation to the ordered performance of work could be directed. In this way, theorizations of control from the mid-nineteenth century onward sought control over that which is fluid and which flows. This control would not be based solely on the constraint of fluidity through channeling, but would rather seek to work from within the very production of fluidity.

The operation of the interface from within fluidity is neither a mixture of smooth and striated nor a shift from one order to the other. Rather, the interface effaces any fundamental opposition of smooth and striated, whether epistemological or ontological, producing them instead as resultant system-states. It is the interface that describes and governs those system-state transformations where, as Deleuze and Guattari describe, "smooth space is constantly being translated, transversed into a striated space; striated space is constantly being reversed, returned to a smooth space."[45] The interface is the space and temporality of translation and reversal; it performs transformation as a fluidity or dynamic form. While as a boundary condition the interface is a kind of striation, it is a striation that is also smooth in that it has already effaced itself within fluid form. Just as fluidity describes the relations that bring the interface into being, so the interface is already given as essential to the nature of the fluid. Fluidity proceeds along a moment-by-moment production and erasure of interfaces; yet the interface also persists as a kind of memory of material fluidity, by which it may be described as dynamic form. Here what is smooth is also at the same time striated.

Control, then, finds its perfect expression in its own erasure. Control begins with a striation, a striation of interface theorized within an interiority that is at the same time the opening up of that interiority to an external imposition of control. The perfection of control is the perfect correspondence of internal state to external imposition, such that all traces of control are perfectly effaced even as control is perfectly exerted. Yet just as the persistence of the interface may be found in dynamic form, so perfect control persists through its erasure, in that the model of control has become perfectly coextensive with the natural materiality that it models. At this limit, that which is controlled is fully embodied by its control. It is as if an act of mimicry were to proceed to identity. The system may then proceed as though it were at once fully natural—as though the act of control had never taken place—and fully controlled.

If control exposes itself as a striation, it is only according to the extent of its imperfection. This exposure, the heavy hand of control, need not be viewed as a failure of control, as control only requires an adequate and not perfect correspondence with that which it controls. In extracting energy from fluid form, a water turbine need not be mistaken for a vortex, even as it mimics aspects of the vortex. Yet while the exposed striations of control, like the vanes of the turbine, may fulfill the tasks at hand, they do not essentially define control. Rather, they point to the gulf between the means and model of control and that which is to be controlled as an opportunity and frontier for further development and refinement.

Information and entropy

The problem of information rose to prominence during the course of World War II. From radar to cryptography, warfare had increasingly become a matter of information processing, servomechanisms, and system design. Information theory, often thought to have been first formulated in a 1948 article by Claude Shannon, applied the methods of fluid dynamics and thermodynamics to the problem of information transmission. Like a dynamic physical system, information flow would be described in a terminology of probabilities, stochastic processes, system states, cycles, equilibrium conditions, and entropy production. However, instead of reactions, pressures, temperatures, volumes, work, and heat, information theory would address transmitters, receivers, channels, messages, signals, and noise.[46] At the same time, work on information would feed back into the study of physical as well as biological systems. Also in 1948, Norbert Wiener proposed "cybernetics" as the name for a new discipline whose aim would be "to find common elements in the functioning of automatic machines and of the human nervous system, and to develop a theory which will cover the entire field of communication and control in machines and living organisms."[47] Flows of information and flows of matter would become increasingly identical in statistical description. Both would be increasingly viewed as behaviors or events and as subject to predictive analysis. What would be asserted in theory as the behavioral equivalence of information and matter, and of animal and machine, was first a program of wartime research; both Shannon and Wiener spent World War II working on antiaircraft fire control systems.

Maxwell's demon was present at the beginning of the shift from thermodynamics to information theory. Leó Szilárd, a physicist known as the central theorist and instigator of the Manhattan Project, laid early groundwork for information theory in a 1929 paper titled "On the Reduction of Entropy in a Thermodynamic System by the Interventions of Intelligent Beings." Against the "very dangerous impression" that Maxwell's demon could circumvent the second law of thermodynamics and allow for the theoretical possibility of a perpetual motion machine, Szilárd argued that measurement is itself quantifiable as entropy. In this way, information would first be described as a thermodynamic quantity, whose production of entropy was "just as great as is necessary for the full compensation" of energy equations within a system. The act of measurement here is inseparable from the measurements themselves; likewise, information must include the activity of its own processing. Here Szilárd accounts for the full system-wide activity of demons—"intellect-possessing beings [*Intellekt besitzendes Wesen*]" who inhabit a "location [*Lage*]" within a system from which to take "measurements [*Messungen*]," and who possess a "memory capacity [*Erinnerungsvermögen*]" that includes storage, retrieval, and erasure. And yet, as much as it may be statistically quantifiable, Maxwell's demon remains a figure of uncertainty and a challenge to statistical description. For Szilárd the demon points to those "quantitative relationships that have not been elucidated."[48] Once again the demon stands for that which in a system remains hidden or unknown.

And as mediated communication takes the form of a complex natural process, the interface again serves as its implicit site of control.

As statistically determined inevitability, entropy is in this way both related and opposed to control. Entropy defines the space within which control operates but also the force it works against. Along these lines, physicist Erwin Schrödinger, in his 1944 treatise *What Is Life?*, defines life as the production of "negative entropy." He identifies this ability as *metabolism*, citing its Greek root *metaballein*, meaning "change" or "exchange." While all living beings are subject to entropy ending in death, metabolism encompasses all those transactions executed against entropy: "Thus a living organism continually increases its entropy—or, as you may say, produces positive entropy—and thus tends to approach the dangerous state of maximum entropy, which is death. It can only keep aloof from it, i.e., alive, by continually drawing from its environment negative entropy.... The essential thing in metabolism is that the organism succeeds in freeing itself from all the entropy it cannot help producing while alive."[49] The activity of life, of metabolism, constitutes a balancing and equilibration of entropies. It is a balancing that has meaning not only in a statistical sense, but also in its performance as a concrete series of transactions, specific in both space and time. Thus in biology it is the *biointerface* that locates the specificity of metabolism.

By 1949 Shannon and computer scientist Warren Weaver would claim the exact equivalence of information and entropy: "The quantity which uniquely meets the natural requirements that one sets up for 'information' turns out to be exactly that which is known in thermodynamics as *entropy*."[50] The notion of information as entropy aligns well with postwar reappraisals of mass media. As in Marshall McLuhan's coinage "the medium is the message," the content of a message would fade in importance to its means of transmission. In media theory, epistemologies of content, which rely on the availability of content as an object of knowledge and interpretation, would increasingly be overshadowed by ontologies of information. Likewise, in information theory "freedom" belongs to the means of transmission. The open channel, free of messages encoded for transmission, possesses the highest degree of freedom, while the filling of that channel with content is a collapsing of freedom and a movement toward entropy and static equilibrium. Likewise, culture, if viewed in any way as an authored text, is increasingly subject to the automatic writing of its transmission. Thus Shannon and Weaver propose a new form of authorship: "Stochastic processes can also be defined which produce a text consisting of a sequence of 'words.'"[51] Differentiations within such a text are subordinate to the differentiations of its propagation. The text of culture is bracketed as the state of minimum freedom within the open channel. It may then be subject to any form of decontextualization or deconstruction in service of the overarching recontextualization that is the possibility of its transmission. What was specific and differentiated within it is leveled out and reformed as an element of fluid flow. Like the "far-reaching liquidation" of history within media described by Walter Benjamin,[52] this becoming-fluid of culture is part of a wide process of capture in which life and society are increasingly unthinkable outside their mobilization within networks.

97

Yet in the growing fluidlikeness of all things, that which impedes flow constitutes both a target and an opportunity. Friction or turbulence within a system may be swept away in smooth flow or exploited in the production of new fluid differentiations. Such differentiations are also opportunities for new interfaces and new potential sites of control. If control over a flow or thing is desired or willed, if that flow or thing is to be controlled, the exertion of control must first negotiate some form of interface. The interface is that which comes into being the moment a model of a system is brought to bear as a constituent means of control. It is the site where the statistical probabilities of the system model must collapse into some critical particularity. Control may only be exerted in the breaking down of the model to a specificity of location and event. In the event, there is no general form of control, as control always involves a specificity of checking against, or targeting, or effecting. Here again the figure of the demon returns as the harbinger of another form of freedom from within the interface as a site of control. The freedom of the demon is not the freedom in potentiality of the open channel or stochastic process;

rather it is the freedom of particularity, of obscurity, of a gaming of the agonistic resolution of opposed pressures, of the atomistic swerve. It is a situated freedom, a constrained freedom, a freedom that only exists in the specificity of its event. It is a semiautonomy, at once obscured and available for programming. It brings agency to control, but an agency that is irreducibly hybridized, accessible only through augmentation.

To culture is given neither the freedom of the open channel nor the freedom of the demon. The first is a negative freedom, a freedom of inactivity under the threat of imminent capture and liquidation within statistical immensity and the speed of information flow. The second is a compromised freedom, an illusory freedom based on an appeal to a particularity that still remains hidden, and which assumes for itself an agency that is only a form of mimicry. As a cultural form the interface bridges the nonfreedoms of the network and the demon, and yet it is also the site for the attempt of culture to reconstitute another form of agency. Here culture finds itself face to face with control, which serves both as the basis of all fluidlikeness and as the moment of its collapse into particularity as the occasion for a controlling, a testing, an extraction of work. As culture once held itself as a fragile bulwark against nature, it now hopes to stand against its own artifice. In this way control becomes a central problem of culture.

Governance and reciprocity

Wiener's coining of *cybernetics* points to the interrelation of control and fluidity: "We have decided to call the entire field of control and communication theory, whether in the machine or in the animal, *Cybernetics*, which we form from the Greek *kubernetes* or *steersman*. In choosing this term we wish to recognize that the first significant paper on feedback mechanisms is an article on governors, which was published by Clerk Maxwell in 1868, and that *governor* is derived from a Latin corruption of *kubernetes*."[53] First referring to the pilot of a ship, *kubernetes* would come to signify a navigation of fluid form. In establishing what Maxwell referred to as "governing" (or alternatively, "regulating" or "moderating") as a direct antecedent of control, Wiener would also point back to the very first coining of the word *cybernetics* in 1834.[54] As part of an over-arching and encyclopedic project to classify all of human knowledge, the physicist and mathematician André-Marie Ampère proposed *cyberné-tique* as the name for the science of governing (*l'art même de gouverner*). Cybernetics would belong to the larger category of politics as one of two branches of "politics in the strict sense [*la politique proprement dite*]"; the other branch of this classification would address the "theory of power [*théorie du pouvoir*]."[55] Just as turbulence for Serres encompasses both the *turbo* of the formless mob and the *turba* of dynamic equilibrium, so the navigation of fluid forms connoted by cybernetics, governance, and control could be seen as extending to political as well as natural conditions.

Wiener's citation of Maxwell's essay "On Governors" brought attention to nineteenth-century concepts of feedback and control that were thought to belong to the twentieth century. For engineer Otto Mayr, "implicitly or explicitly, Maxwell had anticipated a great deal of the conceptual framework of modern feedback control engineering."[56] "On Governors" had largely been forgotten until *Cybernetics*, perhaps for being ahead of its time as much as for its neglect of practical applications and of elaborating its sources and assumptions.[57] If *Cybernetics* described control in terms of circuitry, "On Governors" did so in terms of fluid dynamics. The machines Maxwell cites as examples of governors rely upon the action of water or steam, and his theorization of these machines assumes their operation within a "liquid," such that he describes resistance to machine velocity as "viscosity." A "water-break" invented by Thomson serves a key, though unsourced, reference in "On Governors"; it would only be identi-fied much later as the "centrifugal pump regulator" for a water turbine.[58] In this sense control began with the turbine, which comes to stand as the dynamic form of the vortex now made available for control and the extraction of work. As much as modern control engineering followed from "On Governors," fluid form may be found at the heart of cybernetics.

Maxwell defines *governor* as "a part of a machine by means of which the velocity of the machine is kept nearly uniform, notwithstanding vari-ations in the driving-power or the resistance."[59] The word "governor" had already been used for this class of machines as early as 1863;[60] Maxwell's aim was "the dynamical theory of such governors,"[61] which he formulates as a kind of equilibrium-seeking behavior that Mayr calls "*dynamic stability*."[62] As opposed to the notion of equilibrium as a static or entropic end state, equilibrium as dynamic stability is a continuous balancing in the midst of systemic or environmental variation. While "On Governors" is mostly concerned with the problem of maintaining "constant normal velocity" in the face of "disturbances," Maxwell's work is easily extendable to include the satisfaction of any desired condition that is within the reach of control. In this sense *equilibrium* may refer to any situation in which actual conditions are held in near-enough proximity to a desired state or trajectory through the sustained attention and activity of governance. Equilibrium becomes a state of constant vigilance against disturbance. Likewise, governance has as its end only a local, situated, and provisional equilibration rather than a global or final one. In this equilibration, the purview of governance is to identify and operate upon any discrepancy in a given instance between an actual and a desired state of affairs.

The problem of governance, then, pertains to a discrepancy or difference and the means of its mediation. Forms of governance may subsequently be characterized according to how they mediate discrepancy. For example, in one section of "On Governors," Maxwell distinguishes as "moderators" those machines that, like a simple thermostat, react proportionally to disturbances, and "governors" those that include additional processes to correct errors in reaction time that would otherwise lead to overshooting the "normal value" or desired state.[63] Likewise, discrepancy, whether actual or potential, is a precondition for governance, as well as for "dynamic stability" as the condition produced by governance. What is essential to governance is only the maintenance of the relation it enacts between an existing condition and its own means to alter that condition toward a desired state. As a relation that mediates discrepancy, Maxwell's governance would anticipate the defining mechanism of cybernetics— the feedback loop.

In control theory, a feedback loop defines a relationship where a past state of a system is harnessed in control of its future state. A derivation of the output of a system is fed back into the system as input, forming a loop. As such, the feedback loop encloses a temporal discontinuity. It draws into relation past state and future projection, measurement and desire. It proceeds from one discrete moment of measurement to another, as separated by a cycle of evaluation and response. The gap or discrepancy is as much the delay between measurement and reaction as it is the difference between an event as it occurs and as it is modeled or projected. While control within the feedback loop remains an operation upon a discrepancy, it is no longer primarily an act or event specifically located in space and time, but rather a general condition that defines a system as being under control across a range of possible system states.

Here the loop diagram and the interface may be viewed as distinct modes of representing control. In the loop diagram, control is a general condition or accomplished state. Within the system described by the loop diagram, control is just now occurring, has already occurred, and will continue to occur. The event of its contestation exists abstractly as a set of possible measured and desired values available for calibrated response. In the interface, as the site of control, control is a specific engagement or contested event. Thus Thomson's diagrams of interfaces in fluid flow or Maxwell's diagrams of isothermal interfaces in thermodynamic systems only represent one moment within an ongoing event. The loop diagram, on the other hand, encloses the event of control within the controlled system, essentially representing all possible moments of that event at once as though providing an arena for its contestation. In this way, the interface is implicitly subsumed within the loop diagram. Yet the opposite is also possible; the loop diagram may also be implicitly subsumed within an interface, as on those occasions where a feedback loop is incorporated within an act of control.

As Mayr points out, the concept of the closed feedback loop did not occur to Maxwell, and the mechanisms he describes are more like open, oscillatory systems.[64] This perhaps follows from the role of fluid dynamics in the formation of Maxwell's theory of governors. Through the lens of the interface, Maxwell found in governors an active relationship of opposing forces. Thus governance for Maxwell became the seeking of equilibrium in a contestation, as between velocity and resistance. First abstracted by Maxwell as an equilibration expressed through mathematical functions, the event of contestation would be further abstracted as a loop diagram, rendering it an already accomplished fact, accessible all at once in process and result. Yet the interface remains latent within the loop diagram, from which it emerges in both the breaking down and the enveloping of the general condition in the specificity of the enacted event. Just as the beginning and ending of control is in the act, so the interface describes control's transformation from the specificity of the event to the abstraction of the loop diagram and back again. It is part of the persistence of the interface as a form of relation that it exists in specificity and yet remains implicitly within the general condition. While both the interface and the loop diagram bind elements together into relation, it is the interface within the loop diagram that implicitly performs that relation as an event. This binding together of elements into relation, common to both the interface and the loop diagram, can be termed *reciprocity*.

Shortly before the publication of "On Governors" Maxwell presented a paper at a London mathematical conference titled "On Reciprocal Diagrams in Space and their Relation to Airy's Function of Stress." There he requested help in solving a mathematical problem "in studying the motion of certain governors for regulating machinery," submitting "On Governors" for publication four weeks after receiving the requested advice.[65] "On Reciprocal Diagrams" describes a relation of mathematical functions such that "the first diagram is determined from the second by the same process as the second is determined from the first." In graph theory, reciprocal diagrams are described as possessing *duality*; Delauney triangulation and Voronoi diagrams are an example of reciprocal diagrams. Maxwell found in the reciprocal diagram "a mechanical significance which is capable of extensive applications, from the most elementary graphic methods for calculating the stresses of a roof to the most intricate questions about the internal molecular forces in solid bodies."[66] In static structures, trusses could be analyzed by transforming force diagrams, portraying force vectors meeting at points of equilibrium, into stress diagrams, portraying internal stresses balanced within a closed geometric figure. Here the reciprocal diagram finds within static form a constellation of forces locked in equilibrium. In its transformation it maintains not only the equilibrium of the system but also correspondence with its original diagram, which follows the equivalence by which each is transformed into the other.

Reciprocity, then, is an equivalence through which transformation may occur. Reciprocal elements possess a dual nature as separate expressions bound together across an equivalence of transformation. Thus, in mathematics, the reciprocal of a number is that by which it is multiplied to produce unity. As a condition of mutual determination and an equivalent balancing of forces, the reciprocal diagram is not yet a control diagram. Yet just as the reciprocal diagram of a static structure also describes the potential dynamism should its internal forces be unbalanced, so reciprocity sets the conditions for the introduction of a discrepancy and the possibility of control. In his original 1864 publication on reciprocal diagrams, Maxwell describes as "loose" the reciprocal diagram that no longer holds its forces in equilibrium. In the becoming-dynamic of the reciprocal diagram, "a small disfigurement of the frame may produce infinitely great forces in some of the pieces, or may throw the frame into a loose condition all at once."[67] Likewise, reciprocal diagrams would soon be applied to mechanical linkages such as the Peaucellier linkage, which in an era of steam power was the first planar linkage to translate rotary motion into a straight line. Whether toward a breakdown or an entrainment of force, the reciprocal diagram would first describe dynamic form through an initial static state of forces locked in equilibrium, awaiting disturbance or motive force.

Reciprocity is first a binding of elements together into a unified, mutually communicative system. In binding elements together, reciprocity describes their availability for transformation, measurement, and control. In establishing reciprocity, one diagram, body, process, or field is brought into relation to another such that a space of equivalence is opened up between them. Each is rendered equivalent to the other, as though through an imposition of rules of exchange, so as to draw the two into communication. With the introduction of a discrepancy into its space of equivalence, reciprocity yields a dynamic moment where the action of one element conditions the subsequent action of the other. In this moment the equivalence of reciprocity no longer refers to an equilibrium state or unity but rather to the persistence of a dynamic form.

Reciprocity describes the basic relationship by which the interface comes into being and the means by which it makes a system available to external access. Likewise, the interface could be said to govern out of a condition of reciprocity. While reciprocity does not describe governance, it makes available the means of governance. The interface is the site and performance of a reciprocal game of actions and reactions, in which it constitutes both playing field and rules of play. If control is often assumed to have an essential tendency or teleology toward equilibration and normalization, this is not only in that it first appeared through the lens of equilibrium-seeking machines, but also in that control springs from a moment of equivalence. The interface is the maintenance of that equivalence toward the production of dynamic form. It performs the specific accounting of reciprocity that the loop diagram performs in general. The interface maintains reciprocity as a precondition for the exertion of control over a system, even when that control may drive that system toward instability as well as stability, toward heat as well as work, toward *turba* as well as *turbo*. In this sense control does not require its connotation, common since Maxwell, of being directed toward stability. That which is controlled or under control need not be stable, purpose-driven, bereft of agency, or otherwise confined, except at the moment of equivalence that is the precondition, but not the end, of control. Otherwise control may also express itself in likeness to that to which it is often opposed— the fluid, wild, proliferating, formless, dissipative, tumultuous, or at its limit, free.

The interface and teleology

While cybernetics and control theory would in principle separate the general form of the feedback loop from the particular behavior of

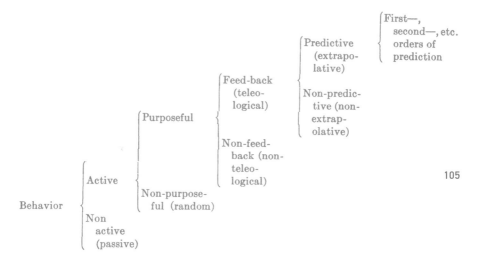

FIGURE 2.4

TELEOLOGY AND CONTROL: A CLASSIFICATION OF BEHAVIOR,
1943. "EACH OF THE DICHOTOMIES ESTABLISHED SINGLES
OUT ARBITRARILY ONE FEATURE, DEEMED INTERESTING,
LEAVING AN AMORPHOUS REMAINDER: THE NON-CLASS....
IT LEADS TO A SINGLING OUT OF THE CLASS OF PREDICTIVE
BEHAVIOR.... IT EMPHASIZES THE CONCEPTS OF BEHAVIOR
AND TELEOLOGY.... FINALLY, IT REVEALS THAT A UNIFORM
BEHAVIORISTIC ANALYSIS IS APPLICABLE TO BOTH
MACHINES AND LIVING ORGANISMS, REGARDLESS OF
THE COMPLEXITY OF THE BEHAVIOR."

SOURCE: ROSENBLUETH, WIENER, AND BIGELOW, "PURPOSE AND
TELEOLOGY," 21, QUOTES ON 22.

equilibrium seeking, the figure of dynamic stability would retain its privileged position in cybernetics in the form of a driving purpose or teleology. This view would be put forward most influentially in a 1943 paper by physician and physiologist Arturo Rosenblueth, Norbert Wiener, and computer scientist Julian Bigelow (hereafter collectively referred to as RWB). As an outline of how cybernetics could be theorized out of communication and control theory, RWB proposed a "behavioristic method of study" that, unlike a functionalist approach, "omits the specific structure and the intrinsic organization of the subject" to focus instead on the *behavior* of that subject, defined in general as "any change of an entity with respect to its surroundings."[68] More specifically, this behavioristic study would track "changes of energy involved in behavior" as the basis for a method of classification that would aim at distinguishing ever more sophisticated types of behavior, specifically focusing on a class of *active* behavior described as *purposeful*. In being active rather than passive, the entity performing the studied behavior is "the source of the output energy involved in a given specific reaction," while in being purposeful, as opposed to purposeless or random, the behavior is essentially "directed to the achievement of a goal."[69] This shift allows a direct comparison of biological organisms and machines according to behavior, especially that deemed "inherently purposeful." Here *servomechanisms* serve as a model for purposeful behavior. Descendants of Maxwell's governors, servomechanisms had become one of the most important lines of World War II research and were the subject of Wiener's wartime work on antiaircraft artillery systems.[70] Thus the first example of a purposeful machine described in the paper is "a torpedo with a target-seeking mechanism."[71]

As a refinement of purposeful, active behavior, RWB proposed *teleological* as a technical term, specifically referring to goal seeking by means of *negative feedback*: "Teleological behavior thus becomes synonymous with behavior controlled by negative feed-back."[72] Unlike positive feedback, which would produce an overall amplification, negative feedback could be used to operate upon a discrepancy to produce a behavior "controlled by the margin of error at which the object starts at a given time with reference to a relatively specific goal." Teleological behavior would then be defined by "a continuous feedback from the goal that modifies and guides the behaving object,"[73] and further characterized according to its predictive power (from nonpredictive to first, second, and further orders of prediction). With examples ranging seamlessly from the organic to the mechanical—from predators chasing prey to machines tracking luminous objects by means of photoelectric cells, from the operation of the human retina to the scanning of a television receiver—teleology would be instrumentalized within a new behaviorism, such that "a uniform behavioral analysis is applicable to both machines and living organisms, regardless of the complexity of behavior."[74] If the term *teleology* once belonged to philosophical discussions of causality or determinism, RWB sought to rescue it as an instrumental term, "quite independent of causality, initial or final," and concerned "merely with an investigation of purpose."[75]

Like Maxwell's demon and reciprocal diagrams, cybernetic teleology was directed against the irreversibility of complex processes. For RWB, "the concept of teleology shares only one thing with the concept of causality: a time axis. But causality implies a one-way, relatively irreversible functional relationship, whereas teleology is concerned with behavior, not with functional relationships."[76] While, like causality, the teleology of a behavior proceeds through time, it does so by drawing on time's axis a space of temporal reversibility in the form of a loop diagram. Like the demon and the governor, the servomechanism finds specificities in time and space within a dynamic condition that would otherwise only be statistically describable, and so opens up the possibility of control. Cybernetics found in teleology its own version of dynamic stability, recapitulating the equilibrium seeking of Maxwell's governor as goal-seeking behavior. The inherently purposeful machine/organism performs as a behavior across time the dynamic fluid form of the vortex. Just as the goal of Maxwell's governor was an equilibration of velocity in the face of disturbance, so the teleology of cybernetics is an equilibration of behavior in the pursuit of a target, as a predator pursues prey. Insofar as this teleology underlies the whole apparatus of information technology, so contemporary notions of communication and information, along with attention and intelligence, are increasingly defined by their goal-directedness. In this way, intelligent technologies evolve within the model space of a predator-prey relation.

Yet the interface, tacitly at work in both governor and servomechanism, does not require such a teleology. The interface attends only to the specific interrelation in which it comes into being. While the activity of the interface may on occasion be described as equilibrium seeking or goal-directed, the end state of the interface is only ever the condition of its own existence. The interface constitutes the zone within and through which the activity of otherwise distinct entities may be translated into an equivalence, through which they may meaningfully and effectively contest, communicate with, or mutually define one another. Through this mediated interrelation, mutually directed formations or actions are made possible, just as the fluid interface operates within the full range of possible fluid behaviors, from laminar flow to turbulence. As much as the interface may be harnessed toward a teleology of control and constitutes the opportunity for that harnessing, it does not yet fully share that teleology. If the interface exerts control, it primarily does so in the maintenance of a reciprocity, a holding together of elements in relation, and only secondarily in the harnessing of those elements brought together by the interface.

If the interface has often been overlooked, misrepresented, or tacitly instrumentalized, it is because of its frequent use as a secondary term in the defining of a primary object. So the interface is used to define the fluid; so the interface, as a condition primarily to be worked through in order to be able to address another object, is tacit to the performance of control. While the equilibrations of the interface resonate with the goal seeking of system-level behavior, the essential difference lies in the performance of equilibration as a precursor rather than an end product of a process of forming or control. As a form of causality the interface recalls Aristotle's efficient cause (*causa efficiens*) more than it does his teleological final cause (*causa finalis*). If the cybernetic teleology proposed by RWB could be extracted from issues of causality by virtue of its behaviorism, then the interface may be viewed as elaborating an *efficiency* of behavior. The teleology of cybernetics, or its target seeking as dynamic stability, is a selective pressure toward those system states from which work may be most efficiently extracted. In this teleology, what is lifelike in machines becomes lifelike in organisms, such that life itself is defined through its availability to operations of harnessing and extraction. In the end, the loop diagram attempts a closed circuit between living and working.

Yet the interface, as a bringing-into-effect and not a goal seeking, exists first as a plane of equivalence and only then as a fluidity of inter-relation. This fluidity may be expressed in laminar flow as much as in turbulent eddy, in work as much as in dissipation, in dynamic stability as much as in instability. Subsequently control, if viewed as a bringing into relation and not as a general condition or end state to be attained, extends to disequilibrium as much as to equilibrium. The interface comes into being as an encounter within each of these cases. At the same time, the encountering of the control interface, in which human beings enter into relation with the machine descendants of cybernetics, takes less the form of an encounter with a lifelikeness than it does an entering into a fluid relation with that lifelikeness. This relation involves a mirroring as much as it does an encounter with a machinic other, since the full en-countering is itself sublimated within a state of augmentation in which the mutual direction of activity may occur. In this way the forming of the interface in the production of fluid form serves to model the forming of the human-machine interface in the production of augmentation. In aug-mentation, as a problem of the human-machine interface, the relation between human and machine proceeds as a kind of fluidity. Just as the interface defines the fluid while remaining tacit within fluid dynamics, so the human-machine interface defines augmentation while remaining tacit in the augmented exertion of control. In both fluidity and augmenta-tion, the forming of the interface describes simultaneous acts of control at different scales: first in the coming into being of the interface, as con-stituent elements are brought into a reciprocal relation; and second in the production of an overall system behavior. Though this first, internal act of control is a kind of equilibration, it is as an existential condition of the interface and not as a teleology. While it is foundational to and tacit within the second, external act of control—what is generally known as control—it is not fully subject to that control, insofar as it is also an encountering of a material nature that is both disciplined and not yet disciplined. Here the interface opens up a window onto forms of fluidity not yet harnessed, or aspects of humanness and machineness not yet subject to augmentation. The interface in this way stands as both prior to and present within the theories of cybernetics.

The turbine as superimposition of fluid and machine

Maxwell begins his discussion in "On Governors" by citing three machines: a hydraulic-brake governor, or "liquid governor," invented by William Siemens; a steam valve governor, invented by James Watt; and a "water-break," invented by Thomson. For Maxwell these machines were typical of most governors in their use of centrifugal force to regulate machine velocity. The water-break was patented by Thomson in 1850 as a "centrifugal pump regulator" and published in a less-known journal in 1851, only a brief excerpt of which is included, as a footnote, in Thomson's collected papers.[77] Maxwell describes its operation: "When the velocity is increased, water is centrifugally pumped up, and overflows with a great velocity, and the work is spent in lifting and communicating this velocity to the water."[78] Here the governor exploits the fluid form of water, subjected to centrifugal force, to produce a braking torque. For engineer A. T. Fuller, who identified Maxwell's citation of the water-break as Thomson's centrifugal pump regulator, the water-break specifically and Thomson's work on fluid dynamics in general was the basis for many of the assumptions used in "On Governors."[79] Fuller further speculates that Maxwell was in part motivated to publish "On Governors" after Siemens's celebrated paper "On Uniform Rotation" (1866) described a hydraulic brake without citing Thomson's water-break.[80] Both machines draw water up through centrifugal motion to slow machine velocity, though Siemens's hydraulic brake does so with a cup-shaped vessel and Thomson's water-break with a forked tube. "On Governors" treats these machines "as if they were equivalent," describing both by drawing upon Thomson's formal treatment of the flow rates of water through tubes.[81]

FIGURE 2.5

VORTEX WATER WHEEL, 1852. "IMMEDIATELY AFTER
BEING INJECTED INTO THE WHEEL-CHAMBER,
THE WATER IS RECEIVED BY THE CURVED RADIAT-
ING PASSAGES OF THE WHEEL.... THE WATER ON
REACHING THE END OF THESE CURVED PASSAGES,
HAVING ALREADY DONE ITS WORK, IS ALLOWED TO
MAKE ITS EXIT." PLAN VIEW OF MECHANISM FOR
A MILL OUTSIDE BELFAST, WITH COVER OF WHEEL
CHAMBER CUT AWAY TO REVEAL ITS INTERIOR.

SOURCE: THOMSON, *COLLECTED PAPERS IN PHYSICS
AND ENGINEERING*, 5, QUOTE ON 4.

FIGURE 2.6

INTERFACE DESIGN C. 1852: THE PARAMETERS FOR
A TURBINE VANE. "*FR* IS THE INNER PORTION
OF THE VANE, AND IN FORMING THE REMAINING
PORTION, ALL THAT NEED BE ATTENDED TO, IS
TO GIVE IT A GENTLE CURVATURE, AND TO MAKE A
SHORT PORTION OF IT AT *S* BE IN THE DIRECTION
OF A RADIUS PASSING THROUGH *S*."

SOURCE: THOMSON, *COLLECTED PAPERS IN PHYSICS
AND ENGINEERING*, 13.

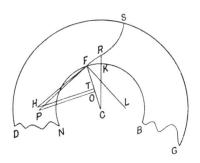

In this sense governance is first expressed as a problem of fluid flow. For Fuller, "Maxwell was interested in problems of fluid flow (not least because of the analogy with electric and magnetic fields) and was doubtless well acquainted with Thomson's early investigations in hydraulics."[82] In the governor, fluidity opened up the possibility of control as an aspect of its materiality and dynamic form, and supplied what was essentially the first carrier of the control signal. Water would become the first true medium of control, and fluidity its first means of articulation. Thomson's invention of the interface as a defining condition of all fluid relations itself came out of his prior work on turbines, in which the vortex would be harnessed in the production of work and the governance of machines.

In this way the machines designed by Thomson in the mid-nineteenth century, including the water-break, may serve as case studies in a genealogy of the interface. The term *interface* would emerge from studies of turbulent flow, and in particular of water subjected to centrifugal force, as an analog of the boundary that separates dynamic stability from instability.

"On the Vortex Water Wheel," a report published by Thomson in 1852, includes as an extended footnote an excerpt of the article he wrote on the water-break.[83] In this excerpt Thomson describes the dimensions, structure, and revolutions per minute of a water wheel within its chamber, given the rate of flow and force of the water. His calculations focus on the vanes within the wheel, against which centrifugally flowing water would transmit its energy in driving the wheel. Along with describing methods to determine the number, thickness, and velocity of these vanes, Thomson includes a diagram by which their form is geometrically derived as a compound of two curves.[84] The geometry of the vane within the vortex water wheel is an abstract diagram of fluid motion. In what Thomson calls its "gentle curvature," the vane approximates the form of the fluid interface within the vortex, instantiated as a machine. The vane interface shapes the fluid flow that passes around it, extracting from that flow energy and work. It is not a channeling as much as a kind of micro-channeling, in that it contributes to the production of the fluid form that it also exploits. In doing so, the vane interface exploits fluid form as if from within its own existential conditions of fluidity. The vane interface reifies the fluid interface whose existence is otherwise only posited as a condition internal to fluidity, by which fluidity is both produced and understood. It is through this reification that the mechanical is embedded within the fluid, just as the vane is the interface by which fluid communicates with machine.

In Thomson's description, the vortex water wheel relies upon the flow of water "through small orifices at high velocity, its inertia being one of the forces essentially involved in the communication of power to the moving part of the mechanism."[85] The communication between fluid flow and machine registered on the water-wheel vane is a translation or transposition of the power and material behavior of fluidity into that of the machine and back again. The overall form of this communication could be described as a superimposition of the machinic turbine and the fluid vortex. Thomson provides an etymology of *turbine* here: "the name Turbine is derived from the Latin word *turbo*, a top, because the wheels to which it is applied almost all spin around a vertical axis, and so bear some considerable resemblance to the top." Within the vortex is superimposed the turbine as a machine for extracting work and the *turbo* of dynamic form: "The whirlpool of water acting within the wheel chamber, being one principal feature of the turbine, leads to the name *Vortex* as a suitable designation for the machine as a whole."[86] Here the vortex describes a state of augmentation with respect to both fluid and machine; from within this augmentation arises the governance of both the coming into being and the dissolution of fluid form. This augmented governance is a seamless integration of self-regulation and the extraction of power. Thomson describes this governance: "In the Vortex, further, a very favorable influence on the regularity of the motion proceeds from the centrifugal force of the water, which, on any increase of the velocity of the wheel, increases, and so checks the water supply; and on any diminution of the velocity of the wheel, diminishes, and so admits the water more freely; thus counteracting, in a great degree, the irregularities of speed arising from variations in the work to be performed."[87] Thomson's vortex is not fully a machine, even though he describes its mechanical articulation.

113

Rather, the vortex is at once the mechanical turbine and the fluid form of the whirlpool, at once an articulation of parts and an irreducible event.

The vortex in this way represents not only a *material intelligence*, once given as essentially unknowable and dangerous, but also, in its super-imposition of fluid and machine, an *augmented intelligence*. Intelligence in this case may be defined as an aspect or quality of a behavior that is at once irreducible and rendered available to control. Thus in cybernetics the availability of intelligence to control would be described by its purpose-fulness, or teleology, as elaborated toward ever greater predictive power. As much as intelligence, material or augmented, is defined by the pos-sibility of control, so intelligence is also defined by the tacit presence of the interface, whether within its materiality or as the means of its aug-mentation. The distinction between a wild materiality and its expression as material intelligence lies precisely in possibility of control and the tacit workings of the interface. To follow a line of development beginning with nineteenth-century studies of fluid motion and thermodynamics, one may find in each of these terms—interface, control, intelligence, augmentation, and, insofar as it is related to these, materiality—an essential fluidity of behavior. If turbulence is, as Serres holds, "an intermit-tence of void and plenitude, of lawful determinism and undeterminism," then it is the space of that intermittence that marks both the site of the interface and the origin of control. It is a space opened up for the inhabita-tion of *daimons*. It is along these lines that all instantiations of the inter-face and of control carry with them a tendency to the vertiginous, to the susceptibility to vertigo that marks the boundary between stability and instability, orientation and disorientation.

The vertiginous moment of interface

Another technical term attributed to Thomson, along with interface, is *torque*, or the moment of force exerted around an axis or in the setting of a body into rotation.[88] The term itself derives from *torques* or *torquis*, the twisted or spiraling metal necklaces once worn as marks of distinction by the ancient Persians, Gauls, Romans, Celts, and others.[89] Like the interface, the concept of torque emerged out of a constellation of ideas

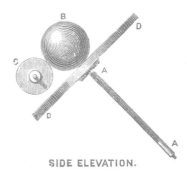

SIDE ELEVATION.

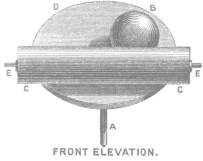

FRONT ELEVATION.

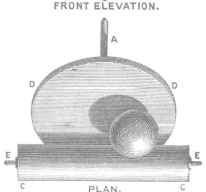

PLAN.

D, the Disk.
A, the Axle of the Disk.
C, the Cylinder.
E E, the Axle or the Journals of
the Cylinder.
B, the Ball.

FIGURE 2.7

"AN INTEGRATING MACHINE HAVING A NEW PRINCIPLE,"
1876. "THIS PRINCIPLE, ON BEING SUGGESTED TO MY
BROTHER AS PERHAPS CAPABLE OF BEING USEFULLY
EMPLOYED TOWARDS THE DEVELOPMENT OF TIDE-
CALCULATING MACHINES WHICH HE HAD BEEN DEVISING,
HAS BEEN FOUND BY HIM TO BE CAPABLE OF BEING
INTRODUCED AND COMBINED IN SEVERAL WAYS TO
PRODUCE IMPORTANT RESULTS."

SOURCE: THOMSON, *COLLECTED PAPERS IN PHYSICS AND
ENGINEERING*, 456, QUOTE ON 454.

connecting fluidity, machine governance, work, and the design of intelligence; again like the interface, torque would feature in the correspondence of Thomson, Kelvin, and Maxwell. Thomson described the use of torque as a method of integrating differential equations in a 1876 paper titled "On an Integrating Machine having a New Kinematic Principle."[90] The resulting machine, used originally to calculate tides and later as the basis of early naval gun fire control systems,[91] prefigured the analog calculating machine widely accepted as the first advanced computing device, the differential analyzer. Developed at MIT in 1931 by Vannevar Bush and his students (Shannon began his graduate studies in 1936 working in Bush's lab on this device), the differential analyzer relied upon a "torque amplifier" to allow the output of one integrator to drive another.

The principles of Thomson's integrating machine were developed between 1861 and 1864, following Maxwell's initial interest and work on the subject. Maxwell's work in turn was inspired by Sang's Planometer, which he saw at the 1851 Great Exhibition in London, and which could quickly and accurately calculate the area of a drawn figure by tracking the motion of a hand-held stylus. In 1855 Maxwell presented a "planometer and integrating machine" to the Scottish Society of the Arts; for Thomson, this device employed "a quite new and very beautiful principle of kinematic action depending on the mutual rolling of two equal spheres, each on the other."[92] Building on Maxwell's work, Thomson developed "a new kinematic method" for an integrating machine "while endeavoring to contrive means for the attainment in meteorological observations of certain integrations in respect to the motions of the wind." He would only publish this work in 1876, after a suggestion by Kelvin that it could be "usefully employed toward the development of tide-calculating machines."[93] Kelvin subsequently addressed Thomson's integrating machine in three short papers, also published in 1876, which describe analog methods of calculating second-order and higher derivations.[94]

The integrating machine consists of a metal disk and a cylinder capable of rotation along separate axes, with the free motion of a metal ball serving to transmit moving force from the disk to the cylinder, and so to perform integrating calculations. The work of calculation was performed through the smooth and continuous transmission of torque from rotating surface to rotating surface. In Thomson's words, "The new principle consists primarily in the transmission of motion from a disk or cone to a cylinder by the intervention of a loose ball ... the pressure being sufficient to give the necessary frictional coherence at each point of rolling contact."[95] Here calculation takes place across mechanical interfaces that come into being as surfaces engage. The torque expressed at the surface of each constituent element of the integrating machine is read and reconciled within such interfaces. Through their protocols, mechanically encoded as contact, position, inertia, and friction, the movements of the constituent parts are bound together as coherent action. The interface is also the means by which the internal coherence of the machine is rendered externally accessible. Thus the integrating machine becomes legible through an additional transmission of work, in which the process of integration is inscribed upon a second "recording" cylinder that draws its power from the first "indicating" cylinder.[96]

Thomson classifies his integrating machine as an *ergometer*—another invented term, from the Greek *ergon*, work—as its function is "measuring mechanical work." For Thomson, "the name 'dynamometer' has been and continues to be in common use for signifying a spring instrument for measuring *force*; but an instrument for registering *work*, being distinct in its nature and object, ought to have a different and more suitable designation."[97] As much as force and motion are harnessed in the production of work, an interface marks the site of that harnessing; likewise, where there is an interface, there is an *ergometer*, a measuring of work. In this working, entities and forces are held in relation so as to possess a coherence that might in Aristotelian terms be called "actuality" (or in a more recent translation, "being-at-work").[98] *Ergon* in Aristotle's usage has been translated as work, function, activity, performance, and defining characteristic; from its root, *erg-*, Aristotle coined *energeia* (actuality, activity), using it almost interchangeably with a subsequently coined word, *entelecheia* (actuality, fulfillment, complete reality). In *Nicomachean Ethics*, Aristotle considers whether "the function [*ergon*] of a human being is activity [*energeia*] of the soul in accordance with reason."[99]

In *Metaphysics*, distinguishing the actuality of form from the potentiality of matter, Aristotle writes: "For the functioning is the end [*telos*], and the actuality the functioning; and that is why the name 'actuality' [*energeia*] is employed with respect to the functioning [*ergon*] and points toward the fulfillment [*entelecheia*]."[100] From within this teleological distinction, *ergon* marks a complication and convergence, in which whatever is form or being may only be so through continuous activity, through constant work and attending. And as much as the interface is the site of this working, it also marks the source from which matter asserts its own actuality, just as it marks the opening up of matter to the fluidities of control.

The integrating machine demonstrates how information technology developed as a *becoming fluid* of machines. As a computer of tides, the integrating machine describes fluidity in both mechanism and application. Torque is seen in this case as an abstraction of the vortex that describes its availability for an extraction of work. At the limit of its abstraction the vortex becomes a surface in rotation upon which is expressed a pure moment of force. The persistence of the vortex as a dynamic form is reformulated as a reciprocity of forces, a coupling of equal and opposite forces across a displacement. This coupling of forces is a holding of forces in place, bounding them to an axis of rotation that marks the center of displacement. It is a reciprocity that describes the *turning* of lines of force, a mutual entrainment of forces in a production of work. It is a harnessing of force toward an integration. In this way torque is a capture of the swerve, or *clinamen*. Such capture represents a step in the transformation of a theory of action into a theory of control. Action is at once entrained as dynamic form and integrated into the machine. Upon the interface are performed both the entrainment and the integration of the act. Insofar as it remains the site upon which such activity takes place, the interface brings into being a kind of fluidity within the machine. This fluidity describes the bridging of action and control; as much as theories of control may seek to subsume the act within the loop diagram, control must still proceed through a fluidity of relation, insofar as its exertion may only take place through an interface.

As a binding to the machine, fluidity is not accomplished through a direct exertion of force; rather, the interface acts only to translate and redirect those forces that seek to work through it. That is, upon the interface power is not exerted directly onto its object, but rather is diverted; the interface comes into being with a turning, a bending of force into torque, a swerve. The interface, whether hypothetically posited or in action imposed, is the site of this turning. It is through such a turning that the particular is exposed within the general, and the interface comes into being as a relation of particularities rather than of generalities. It is in this way that the interface comes to inhabit the full measure of an action in its every operable moment in space and time. Here the power of the interface may be compared to Bachelard's description of the productive, elemental power of water: "This duration is a substantial becoming, a becoming from within."[101] Such a becoming proceeds from the coming into being of the interface in equipoise, yet in an equipoise situated upon the turbulent boundary of turmoil and order from whence control originates. Consequently, with respect to the interface, even the straight line carries within itself at each moment of its calculation the possibility of vertigo.

This is also to suggest that human-machine interaction constitutes a kind of fluidity or fluidlikeness. Just as modern conceptions of control first emerged from within the context of fluid dynamics, so fluidity aptly describes the mutual entrainment or reciprocity that holds in the relation of human beings to and through machines. Where "interactivity" describes an existentially disengaged relationship, in which entities communicate across an uncontested space and without assuming the risk of capture and transformation, "fluidity" describes a complex form of engagement, in which constituent entities are bound into relation in the mutual production of a dynamic form. It is in a kind of fluidity that the machinic and the biological are bound together, and it is through this binding together that both entities, facing each other across the interface, may produce in their mutual activity a fluidlikeness that may also become a lifelikeness. Here, in the zone of contact of this binding, the interface serves as site of contestations and resolutions, of reciprocities and *daimonic* influences, of communication and control; from within the interface comes the *apparition* (as from the Latin *apparatus*, from *apparo*, to prepare) of the ghost in the machine.

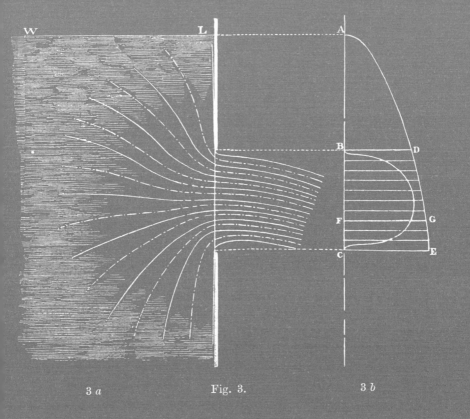

THE AUGMENTATION OF THE INTERFACE

A genius of augmentation

The Roman god Genius was held to encompass the whole nature of a person as both created and creating. Often associated with the forehead, Genius held claim over human agency and will. Dumézil identifies Genius with its Latin derivations *ingignere*, "to cause to be born in," as well as *ingenium*, the nature of that which is engendered.[1] It refers to both an engendering and what is engendered, whether as innate quality or natural disposition, constitution or character, mental ability or power, invention or ingenuity, or even as a trick or clever device. Its engendering both precedes agency and conditions it; it includes as well the formation of the political subject. For Dumézil, "well before the idea of 'person' was clearly distinguished in law, it was Genius, in religion, which approached it most closely."[2] Genius would also be "attributed to the 'moral persons' constituted by families, the state, the provinces, the colleges, and military units," as "the expression of the originality, of the distinctive personality, and occasionally of the esprit de corps of these various collective bodies." Likewise places were also considered to possess a Genius, as with the expression *genius loci*, or spirit of place; this notion "proliferated to the point where the gates, the baths, the market places, even the smallest *anguli* all produced their own Genii."[3] Thus Genius would overlap with the guardian gods of boundaries, or *Lares*, and in particular the *Lares compitales* as gods of crossroads, who inhabit the liminal space where property lines meet.

If Genius engendered identity or personhood, it also represented those impersonal or exterior forces by which engendering took place.

Thus it easily extends from persons and groups to things and places. For Agamben the self or subject emerges from a continuous dialectic between Genius and Ego, between the impersonal, uncontrolled, and innate, and the personal, controlled, and conscious elements of selfhood. These constituents of the self "coexist, intersect, separate, but can neither emancipate themselves completely from each other nor identify with each other perfectly."[4] He further identifies Genius with both the urge to create and biological processes that maintain the body: "It is Genius that we obscurely sense in the intimacy of our physiological life, in which what is most one's own is also strange and impersonal, and in which what is nearest somehow remains distant and escapes mastery." The poetic is then "the life that maintains the tension between the personal and the impersonal, between Ego and Genius."[5]

The problem of Genius may also be viewed as a problem of augmentation. The state of augmentation constitutes an inherent challenge to selfhood, in which the subject takes on an externality, technological or otherwise, as internal to itself in seeking claim over the hybrid agency that emerges from augmentation. Across the human-machine interface, Genius corresponds precisely with *intelligence*, insofar as it serves as the means of augmentation—defining the protocols by which human and machine are brought into alignment—and as an end of augmentation, insofar as intelligence describes the criteria of augmented behavior. Like Genius, intelligence describes an encounter between the human or personal and the nonhuman or impersonal. Thus intelligence is more and more identified with machines, material or biological systems, or even organizations; at the same time, intelligence extends from conscious acts of cognition into the preconscious, tacit, and embodied.

Intelligence becomes a forging of connections, whether between conscious and tacit or human and machine. This corresponds with the *ingenium* described by eighteenth-century philosopher Giambattista Vico, used in opposition to Decartes's logical method, and often translated as "wit" or "mother wit." For Vico, "*ingenium* is the faculty that connects disparate and diverse things"; as a human faculty it is "the creative power through which man is capable of recognizing likenesses and making them himself."[6] Here Vico's characterization of human reason and thought immediately precludes solipsism or a Cartesian separation of mind and body, in that *ingenium* is already given as engaged in making or producing. In this way *ingenium* is an embodied intelligence, one that as "mother wit" is literally birthed into humanity. The challenging of the self in augmentation might be expressed as an occupation of human *ingenium* by a *Genius* of technology: a haunting of the self by those impersonal forces, imparted across the interface, that have produced it. As to the location of this challenge within human agency and experience, we might we ask, following Vico: "Is it because, just as nature generates physical things, so human *ingenium* gives birth to mechanics, and as God is nature's artificer, so man is the god of artifacts?"[7] If a god, then only most tenuously so; it is the uncanny but *genial* relation with machine intelligence that describes the conjoined attraction and suspicion that it engenders, and equally a sense that the boundaries between the real and the simulated may no longer hold. For insofar as intelligence upon the interface is material and technological as much as human, it bears less relation to conscious thought than it does to the preconsciousness of its embodiment.

The tacit knowing of the interface

In its drawing together of the capabilities of human and machine, the interface operates as a threshold condition through which both knowing and acting are enabled. An interface enables only as it is worked through in the production of a state of augmentation. In the experience of the user the interface takes on a seeming transparency as it is worked through, and as its user is enabled through augmentation. Here the threshold is attenuated toward an apparent disappearance. The experience of the interface as a form of subjectification takes on an illusory quality. The immediate encounter of the interface as a form of separation, as a thing that challenges, that must be attended to, and is not yet but soon will be available for use, blurs and fades into the experience of the interface as a form of augmentation, as a thing that has already been worked through and so no longer appears as an object of attention. That the interface should seemingly disappear in use follows its mode of operation. This occurs not only in the sense that a mirror seems to disappear in becoming the image of what it reflects. More significantly, the user sublimates a real experience of the interface as form of separation into an enacted experience of the interface as a form of augmentation. The user attends and acts through the interface, and assumes dominion over the state of augmentation that enables the attending and willing. Yet the actual working through of the interface, and the continued state of separation that persists within the moment of augmentation, remains, despite assumptions of its disappearance.

In part, the disappearance of the interface is illumined by two frequently addressed and related topics in twentieth-century philosophy and psychology—how objects or tools become ways of seeing and knowing, and how certain complex techniques, once learned, may be performed without requiring conscious attention while still contributing to consciously attended performance. Michael Polanyi described this latter situation—possessing and using knowledge without conscious awareness of its possession or use—as *tacit knowing*. Arguably, most of human knowledge is tacit or subliminal; this would include forms of knowledge that are in some way embodied. Yet insofar as tacit or embodied knowledge often resists formal description or remains hidden from direct observation or introspection, it is often neglected as a form of knowledge. Polanyi proposes an axiom: "We can know more than we can tell." This is especially the case when tacit knowledge forms the basis for other kinds of knowing, as when "in an act of knowing we *attend from* something for attending *to* something else; namely, *from* the first term *to* the second term of the tacit relation."[8] For example, the conscious performance of a skill (the second term, attended to) might rely upon a sequence of kinesthetic movements that have been trained in repetition to the point where conscious attention is no longer required in their performance (the first term, attended from). In this way the form of knowing and acting that is attended from is hidden within or subliminal to the form of knowing and acting that is attended to.

Our awareness of tacit knowledge is often indirect and gained through inference; it may be expressed in functional terms, identified as a requirement for another form of knowledge or action, or in phenomenal terms, reflected within in the "appearance" of what is known. That tacit knowing should find itself reflected in its object demonstrates how it structures that which it attends to according to its own form. In Polanyi's words, this is a "correspondence between the structure of comprehension and the structure of the comprehensive entity which is its object."[9] If knowledge is such a comprehension—that is, if it possesses a form directed toward a comprehensiveness or coherence, whether in the knowing of a thing toward a completeness, or in an action aimed toward a fulfillment—then what is known tacitly describes the particularities and fragmentations that underlie the comprehension, just as an inexpressible tacit knowing underlies knowledge that is expressed. The structure of knowing and acting, which proceeds from subliminal to conscious attention, is reflected in the structural coherence of that which is known and acted upon. In this way the crossing of a threshold of awareness is inscribed within any consciously directed performance. Thus, for Polanyi, "tacit knowing of a coherent entity relies on our awareness of the particulars of the entity for attending to it; and if we switch our attention to the particulars, this function of the particulars is canceled and we lose sight of the entity to which we had attended."[10] Tacit knowing structures a knowledge within which it otherwise remains separate and inaccessible, whether as underlying criteria of that knowledge or in its adherence to that which is thereby known.

125

Yet even if tacit knowing resists expression within given categories of knowledge, as Polanyi describes, it may nonetheless be opened up by another kind of knowing. This is an augmented knowing, in which human ability, conscious and subliminal, is augmented by tools, equipment, or techniques to constitute a way of knowing, measuring, and structuring what would otherwise remain tacit in human sensation, cognition, and action. With the advent of artificial intelligences and human-machine interfaces, tacit knowing becomes a primary site of intervention for the emergence of hybridized forms of knowing that are both human and nonhuman. Such hybrid forms of knowing are both essentially alien and intimately familiar to human understanding: alien in incorporating forms of knowing that follow a machinic or nonbiological evolutionary path; and familiar in that this evolutionary path also tracks a lineage of desire expressed in human design and use.

This is to propose a linkage between tacit knowing and its effects both above and below the threshold of conscious attention, and the evolution of processes of augmentation. Here augmentation not only operates through a kind of tacit relation, as one may work through an interface to perform a task, but also opens up as territory to be developed that which was once tacit or hidden in human capability. Augmentation develops this territory according to the domains at its disposal—social and material, political and technological—and by means of processes of subjectification that are to a large part subliminal to the attention of their subject. It addresses the ways artifacts play upon subjective experience, including the production of an augmented subjectivity. Within augmentation the subject of technology and the subject of politics meet. As conscious and unconscious attention is increasingly modified and channeled through an internalization of devices and machines, augmentation holds ever greater sway over human life. If tacit knowing is non- or prerational, primordial with respect to its availability to knowledge, it also describes the territory available to another form of rationalization—hybrid, machinic, and augmented. In this way tacit knowing marks the site of a potential augmentation.

The tacit may also be found in what Heidegger has called *readiness-to-hand* (*Zuhandenheit*), or "the quality of being at our disposal."[11] For Heidegger, readiness-to-hand denotes an ontological category or way of being in the world "in which entities as they are 'in themselves' are defined." It is found in the use of tools and equipment, in the sense that equipment in its use "has its own kind of sight, by which its manipulation is guided."[12] When such equipment is removed from the context of its use, it is no longer ready-to-hand but has been broken down into another category. *Unready-to-hand* includes equipment that requires repair or whose user does not possess the knowledge of its use, such that the unready-to-

126

hand is encountered as an obstacle or problem that requires solution; while *presence-to-hand (Vorhandenheit)* describes equipment that has been rendered available to a rational analysis that first strips it from the context of its use.

For Heidegger, presence-to-hand denotes the means by which modern science lays full claim to the entity it analyzes, granting that entity a fully immanent presence and availability to description; it thus precludes a notion of meaning that emerges through use over time. Heidegger identifies in presence-to-hand a worldview that accompanies modern science and technological development, in which knowledge is privileged as rational insofar as it has been abstracted from life in the production of a coherence. In its abstraction and coherence, the world in its actuality is concealed and leveled. The knowledge of things "in themselves" or as they are in the world is hidden within a knowing that belongs to engagement and use, a knowing that might also be termed tacit. Heidegger here proposes a moment when "the world is lit up [*aufleuchten*],"[13] when that which has been in use as ready-to-hand just begins to fall out of use, whether toward a breaking down into disuse or toward being "disclosed" as present-to-hand. At this moment "the worldly character of what is within-the-world [is] lit up," in that the nature of the thing in its use is exposed not as an obstacle or abstraction, but according to its identity and creative force. The exposure of the creative act, whether in the activity of the artisan's workshop or embodied in the human sensation or cognition, is in itself disruptive to received categories of knowledge: "The presence-at-hand of entities is thrust to the fore by the possible breaks in that referential totality in which circumspection 'operates.'"[14] Here a totalizing ontology of presence is contested within a moment in which the world reveals its nature to human experience according to its engagement with things in the world.

Yet it is also along these lines that augmentation brings with it a new phenomenology of experience. The evolution of the human-machine interface is directed precisely toward the augmentation of processes of sensation and cognition that were once fully embodied and concealed within either human or machine. In this regard the phenomenology of perception of Maurice Merleau-Ponty also describes a kind of augmentation, in which objects are first perceived and then perceived through. With the *inhabiting* of one object, one may *grasp* others: "To look at an object is to inhabit it [*regarder un objet, c'est venir l'habiter*], and from this habitation to grasp [*saisir*] all things in terms of the aspect which they present to it. But insofar as I see those things too, they remain abodes open to my gaze, and, being potentially lodged in them, I already perceive from various angles the central object of my present vision. Thus every object is the mirror of all others."[15] Here the object entails a subjective relation that is both static and active; it becomes active in a seizing or grasping that is first enabled by the stasis of an inhabitation. Just as the static and the active are bound together within a single perception, so are the subjectivity of the perceiver and the objecthood of that which is perceived. It is not only the object that is mirrored in perception but also the perceiver. Following Polanyi, both the inhabitation of the first object and the reaching from within that inhabitation to grasp a second object could be viewed as a tacit knowing. Only the second object, upon having been grasped, would then possess the coherence of being consciously knowable, even as much as this coherence is founded upon the tacit knowing that structured it and brought it into being. Such is the case with the interface, as a condition that is similarly inhabited at the same time that it is worked through, and that tacitly structures the objects of its perception or action. Yet instead of finding resolution in the perception of a human perceiver, the interface describes a kind of a phenomenology of augmentation in which resolution occurs in a hybrid perceiver, or augmented subject, within a state of augmentation.

Likewise, the interface is not reducible to a technology of immanence. This is the case even as its operation as a form of separation would seem to conceal itself in the production of an augmented view of the world in "real time," or so as to appear to instantaneous. Though the interface may also be described as ready-to-hand, it opens up an experience once embodied in human use to a hybridized embodying. While this human-machine embodiment is partially the product of human artifice, it is equally the product of the technological lineages it incorporates, which evolve according to human use and to the inherent qualities of matter and machine. Here the interface both models and develops its human use as a territory for expansion, supplementing attention and action above and below the liminal threshold of conscious awareness, toward a common trajectory with the machine in the production of a human-machine system. In this sense the interface does not seek the abstraction of human abilities and potentials, even as it may on occasion work through such an abstracting. Rather, the interface determines its own context, within which actions may come into being, a context that does not rely upon a metaphysical structuring or rationalized abstraction to the extent that the interface is at least in part free from the constraint of human categories of knowledge. As hybrid condition the interface traces its own evolution of knowing and acting, and so remains essentially indeterminate. If the full expression of the interface springs from the state of augmentation it produces, then the state of separation, and its production of the fragmented subject, stands as its tacit knowing.

Singularity

In the first, postwar decades of AI research, advocates of "strong AI" elided the distinction between human and machine intelligence by constraining both to problems of abstract problem solving. In *What Computers Can't Do* (1972), philosopher Hubert L. Dreyfus would cite Polanyi, Merleau-Ponty, and Heidegger in his influential critique of the ontological claims of early artificial intelligence, a critique that now reads like a forecast of the evolutionary path machine intelligence would soon adopt. Illustrative here is his "classification of intelligent activities," progressing in complexity from *associationistic* activities that are "innate or learned by repetition" and are programmed as decision trees or templates; to *simple formal* activities that are "learned by rule" and programmed as algorithms; *complex formal* activities that are "learned by rule and practice" and programmed as "search-pruning heuristics"; and *nonformal* activities that are "learned by perspicuous examples" and are essentially unprogrammable.[16] This last category is a kind of tacit knowing that Dreyfus describes as intuitive and "open-structured," and of "an entirely different order" than the other three areas: "Far from being more complex, it is really more primitive, being evolutionarily, ontogenetically, and phenomenologically prior to [the other areas], just as natural language is prior to mathematics."[17] For Dreyfus, early AI research failed to appreciate the qualitative differences between these areas, such that early success in programming activities of the associationistic and simple formal types led to an assumption that all aspects of human intelligence would soon be machine programmable. This would carry with it an implicit danger: "If the computer paradigm becomes so strong that people begin to think of themselves as digital devices on the model of work in artificial intelligence, then, since ... machines cannot be like human beings, human beings may become progressively like machines."[18]

In Dreyfus's argument, becoming machinelike is an entrapment within a circular argument, which first identifies within human cognition a behavior available to machine computation and then projects that machine computation back onto human cognition. Yet as much as Dreyfus's critique was borne out in the failures of early AI, and as much as AI has since evolved, intelligence remains a site of contestation between human and machine. Having discarded the explicit indexicalities of early AI, intelligence-as-contestation would expand its territory into the nonformal, the self-organizing, the embodied, and the tacit. The expanded field of artificial intelligence—which encompasses agent-environment interactions, behavior-based robotics, parallel-processing architectures, network protocols, high-level coding languages, gaming and social media computing environments, and more—finds as its frontier the territory opened up by interfaces of various kinds. Just as the assimilation of postwar command economies into a networked global market seemed to disperse and naturalize, but nonetheless ultimately enhanced, the hold of economic power over life, so the dispersal of intelligence into incalculable contestations and resolutions enacted across myriad interfaces has described the gradual remaking of human life and society as a hybrid condition.

Along these lines, postwar arguments for strong AI are less important as truth claims than as efforts to shape the outlines of a battlefield as it comes into being. In 1958, for example, Herbert Simon and Allen Newell argued that human intelligence proceeds through heuristics, or rules of thumb, and then cited programs of their own design as examples: Logic Theorist (1956), often credited as the first AI program, and General Problem Solver (1957). Their conclusion is cited by Dreyfus: "Intuition, insight, and learning are no longer the exclusive possessions of humans: any large high-speed computer can be programmed to exhibit them also."[19] Despite the overreach of this claim, heuristics—as a form of structured problem solving that circumvents exhaustive searches for optimal solutions to search out more quickly gained sufficient approximations—has subsequently become ubiquitous and essential within the human experience of technology.

Heuristics is as much a computational efficiency as it is an attempt to make sense of the scope and demands of programming. In this way it delimits a testing ground where human notions of organization and process meet with machine performance. Both Simon and Newell had moved from organizational theory into AI, and Logic Theorist was first developed by an organized group of people processing instructions written on index cards, which described subroutines and memory, before it was programmed into a computer.[20] Coding would soon become a multi-layered process, from machine code to assembly code to high-level languages to operating systems, each layer of which is a further separation from a machine event and a further opportunity for human practices, preferences, and assumptions to obscure what properly belongs to the machine. Already by 1958 Oliver Selfridge had proposed that problems of tacit knowing such as pattern recognition be addressed through a bottom-up, parallel-processing computer architecture he called Pandemonium; here information would flow up a layered hierarchy of semiautonomous subroutines or *demons* in the course of producing an overall decision.[21] Coding might then be viewed as played out within the space of a fluid hierarchy, with each moment of decision describing both a contestation and a mutual mirroring of human and machine. It is within this space of uncertain hierarchy, perhaps, that Wendy Hui Kyong Chun identifies software as a form of ideology, in its attempt to "map the material effects of the immaterial and to posit the immaterial through visible cues."[22] Where postwar AI laid claim over human intelligence through a supposed equivalence to abstract symbol manipulation, software in its use describes an environment in which intelligence is sought out within an accretion of self-reflective techniques, sites, and desires.

Related to the claims of strong AI, the *technological singularity* recasts the claims of machines over human intelligence as a coming event—an irreversible future moment when machine overtakes human as paradigmatic of intelligence.[23] And yet the singularity is less interesting as futurological conceit than as a designation of present contestations. The tie between singularity and contestation might begin with its original use by mathematician and polymath John von Neumann, whose experience with contestation included foundational contributions to game theory and the physics of the hydrogen bomb. Writing after the death of von Neumann in 1957, the mathematician Stanislaw Ulam, a close collaborator in the Manhattan Project, paraphrases von Neumann's speculation on how "the interests of humanity may change, the present curiosities in science may cease, and entirely different things may occupy the human mind in the future." As an example, Ulam relates how "one conversation centered on the ever-accelerating progress of technology and changes in the mode of human life, which gives the appearance of approaching some essential singularity in the history of the race beyond which human affairs, as we know them, could not continue."[24]

The singularity describes a moment where intelligence—along with systems of meaning in general—ceases to make reference back to what is human in order to follow instead a technological or hybrid lineage. In response, humanness would cross another threshold of transformation to become something beyond, or other, than human. Yet with this erasure is a kind of relief, an unburdening of those inexpressible and irreducible aspects of humanity that could find no technological expression. If the human relation to technology is a kind of perpetual becoming, a project thrown into the future as an endless process of refinement, the singularity represents a full attainment of being, in which humanness finally surrenders its claim over being in the world to the machine. It imparts to technological development both a grand narrative and an end of all narratives, a stepping out into the unknown. Thus singularity is often argued as a mixture of scientific discourse and science fiction speculation.

Yet the singularity is also a present contestation. It is perpetually rehearsed within each moment of augmentation. It embeds within the immediate experience of augmentation a future moment in which a pitched struggle reaches its breaking point—a moment of utmost criticality and almost complete opacity. It can do so since the future moment is also fully within the present as a moment of pure contestation. What is contested is reconciliation, a rendering-equivalent that would allow for a communication. The singularity, then, is also the coming into being of augmentation at every instance. It is a present testing, a probing of defenses, a marshaling of forces, that also carries within it the promise of an ecstatic moment of absolute cessation and unconditional surrender.

In its pure contestation, the singularity describes the extent to which the relations of humans and machines are essentially atopic, stripped of specificities of place, context, and occurrence. This is so even as it also foregrounds these specificities as so many possible tactical sorties in an ongoing and unending struggle. Likewise, the positive in singularity may just as easily be negative; it may be utopian just as easily as dystopian. Thus the ease with which contemporary culture can imagine its relation with technology as a smooth blurring of the real into the virtual; and thus those science fictions where what was thought to be simulation is discovered to be real, or where what was thought to be real turns out to be simulated, or where the simulated real becomes preferable to the real. This in turn follows the perpetually unfulfilled need of a subject only partially apparent in fragmented form in the face of technology to merge seamlessly with the spectral completeness of an augmented other. In this sense life, or being in the world, becomes through its encounter with technology the object of a particular combination of hope and suspicion, of exhilaration and suppression. Insofar as life becomes available to a kind of simulation in which every detail is indistinguishable from the real, so technological development carries the promise of a complete alleviation of all aspects of humanness that appear as a weight or inertia unto immobility, with respect to which technology represents a seeming weightlessness and mobility; yet with this assumed state of weightlessness comes also a vertigo of suspicion that another agenda or entrapment may also be at work, and that the bargain made with technology has always been a Faustian one.

If such a cultural tension is granted to exist between the virtual and the real, then the figure of the interface, both as a concept and as a collection of ways in which human beings singly and socially interact with and through machines, delimits the site of that tension, the arena within which the human and the machinic contest one another with respect to both the simulated and the real. In particular, the interface emerges as essential to the governance of the transitions and transactions that describe both the encountering of technology as a fragmentation and its use as an augmentation. As a site of encounter, the interface demonstrates that no encountering of technology may be ultimately viewed as siteless or atopic; in this sense the interface is also the end of singularity, the end of augmentation, the threshold across which the contestants are fully separated in order to begin the struggle anew. As a state of relations that begins with and operates through a separation, the interface demonstrates that no passage into augmentation may reach permanence and fulfillment as long as the interface remains. For even if the interface suggests in its provision of augmentation the possibility of leaving behind once and for all the weight and drag of humanness, it also operates continually upon that very humanness, dividing it, shaping it, and bringing it to light, finding within it a territory of limitless expansion. The history of the interface is in this way a history of intelligences and lifelikenesses as well, insofar as the interface delimits the boundary condition across which intelligences are brought into a common expression so as to be tested, demonstrated, reconciled, and distributed, and through which new forms of lifelikeness are produced and experienced. And just as the singularity upon the interface is an immanent moment of equipoise between separation and augmentation, so the future technological singularity describes the endgame of the interface, the threshold beyond which the condition denoted by interface ceases to exist.

Symbiosis

In his preface to *What Computers Can't Do*, computer scientist (and later influential defense advisor) Anthony G. Oettinger relates how "Dreyfus's own philosophical arguments lead him to see digital computers as limited not so much by being mindless, as by having no *body*."[25] Oettinger had already by 1963 run up against the limitations of existing AI while working on automatic language translation, a then heavily funded field meant to have been one of AI's premier applications. His defense of Dreyfus would be echoed by psychologist and computer scientist J. C. R. Licklider, an early advocate and theorist of human-computer interaction (HCI), who like Dreyfus supported the idea of "man-machine cooperation."[26] Licklider's work on computer networks (including ARPAnet, SAGE, and Project MAC) dovetailed with his theorization of "Man-Computer Symbiosis," as in a widely cited article of the same name published in 1960. Here he imagined the possibility of a human-machine relationship as intimate as that between the fig tree and fig wasp: a coevolutional mutual dependence in which the fig tree relies solely upon the fig wasp for pollination and the fig wasp relies solely upon the fig tree as the site of its reproduction. Citing a dictionary definition of symbiosis as the "living together in intimate association, or even close union, of two dissimilar organisms," Licklider hoped that "in not too many years, human brains and computing machines will be coupled together very tightly and that the resulting partnership will think as no human brain has ever thought and process data in a way not approached by the information-handling machines we know today." He would thus happily posit a singularity in which human-machine symbiosis becomes a dominant form of intelligence, defining an era that "should be intellectually the most creative and exciting in the history of mankind."[27]

In part Licklider's interest in symbiosis followed from his position bridging two disciplines, computer science and human factors and ergonomics (HFE). Symbiosis would be posited as a subclass of human-machine systems, then a focal point of HFE, in a journal addressing human factors in electronics. HFE came to prominence in World War I, where it drew primarily upon psychology and scientific management to address the selection and training of personnel and the design of equipment and environments for human use.[28] Given the complexity of the problems of interaction between humans and machines or organizations, HFE would from the beginning rely on a cross-disciplinary and multiscalar approach, one that would explicitly precede and influence the field of HCI.[29]

HCI is in many ways a child of World War II, which drove the shift from analog to digital, foregrounded the problem of information in logistics and organization, and conditioned the first interactions between human and digital computer. Among the first figures of HCI is Vannevar Bush, an MIT engineer who, as director of the Office of Scientific Research and Development, oversaw the management of the entire US wartime research effort. Before the war Bush had headed the design team that produced the first general-purpose analog computer, the differential analyzer (a device that followed in principle James Thomson's integrating machine). Immediately at the end of the war Bush would write "As We May Think," an article aimed toward the popular press that identified the technological augmentation of intelligence—learning, information production and retrieval, and so on—as the most central issue facing science. Among the hypothetical devices outlined in this article, the best known is the *memex*, "a sort of mechanized private file and library" that would mirror and augment the human capacity for "selection by association." Anticipating hypertext, Bush described its operation: "With one item in its grasp, it snaps instantly to the next that is suggested by the association of thoughts, in accordance with some intricate web of trails carried by the cells of the brain." The *memex* would unite the indelible memory of a file system with "the speed of action, the intricacy of trails, the detail of mental pictures" of human cognition.[30]

One might also include as a significant event in early HCI a hypothetical test devised by Alan Turing. On the eve of the war Turing had established the theoretical basis for digital computing in what would be called the Turing machine. (His work during the war is now widely acknowledged as the greatest contribution to Allied cryptography, and contributed to an effort that in 1943 yielded Colossus, the first programmable digital computer.) The Turing test, as described in an essay of 1950, addresses the question of machine intelligence by positing a game of determining which of two remote and unseen interlocutors is male and which is female, then substituting one of these with a machine designed to give humanlike responses. Turing then asks: "Will the interrogator decide wrongly as often when the game is played like this as he does when the game is played between a man and a woman? These questions replace our original, 'Can machines think?'"[31] Here Turing defines intelligence fully in terms of a relation or interaction, and what is more, a relation mediated by some form of remote communication. The question of an essential, qualitative similarity between the intelligences of human and machine is both delayed by reference to a future event and delimited through an interactive game of imitation and mimicry.

In the postwar era of HCI, Licklider and others would explore the integration and symbiosis between human and machine both speculatively and in the design of artifacts. Computer scientist Douglas Engelbart, writing in 1963, several years before patenting the computer mouse, describes the agenda of his work as "augmenting man's intellect" and thereby bringing about "a way of life in an integrated domain where hunches, cut-and-try, intangibles, and the human 'feel for a situation' usefully coexist with powerful concepts, streamlined technology and notation, sophisticated methods and high-powered electronic aids."[32] Architect-turned-information-technologist Nicholas Negroponte would write in 1970 of a computer-based "architecture machine" where both designer and machine "track the other's design maneuvers" in a "progressive intimate association of the two dissimilar species" that "evolves through mutual training" and "has no historical precedent"; further, "with direct, fluid, and natural man-machine discourse, the former barriers between architecture and computing would be removed."[33]

A 1983 book, *The Psychology of Human-Computer Interaction*, would find represented among its authors expertise in psychology, computer engineering, AI, and design; it would identify the central interest of the HCI as the production of a dialogue where "both the computer and the user have access to a stream of symbols flowing back and forth to accomplish the communication; each can interrupt, query, and correct the communication at various points in the process."[34] *Interface* for these authors denotes "all the mechanisms used in this dialogue," and in particular, as objects of a design process, "the physical devices, such as keyboards and displays, as well as a computer's programs for controlling the interface."[35] Yet here Dreyfus's objection returns as to the form and context of such dialogues and their technological means. If the interface is solely an issue of technological design, even if informed by aspects of human behavior, what else is entailed within an ensuing human relation with technology that is not only a reflection of but also an imposition upon humanness? Even if, to borrow Licklider's analogy, the fig wasp is, in its symbiotic relationship, not in danger of what Caillois called *psychasthenia*, where through mimicry it imagines itself to be the fig tree, it nonetheless conforms itself to the fig tree in both morphology and behavior. The extent of this conformance remains invisible so long as the interface is viewed as confined within a design problem, and thus is subject to a solution, no matter how provisional such a solution is taken to be.

If HCI today is less concerned with human-machine symbiosis as a single defining goal, it is only because its aims and techniques— addressed within the discipline in the rise of ambient intelligence, ubiquitous computing, and *everyware*—have spread out into all available sites, scales, and aspects of life. Each of these scales of effect may be populated by its own moment of symbiosis, its own augmented performance of dialogue, integration, feeling, intimacy, mutuality, fluidity, and so on. Its available means would be diverse, encompassing industrial or environmental design methodologies, usability-testing techniques, materials sciences, artifactual interface designs, algorithms, AI (from agent-environment interactions to artificial life), networks, robotics, prosthetics, bionics, and more. While any given HCI project may be viewed according to given specifications of technology and use, the overall trajectory defined by the interface across society and life, of which HCI is part, is one of an evolving of relations at scales of ever greater intimacy and differentiation.

138

Here the interface points to the relation by which it comes into being. Its historical development is less a design issue than it is a coevolutionary process of contestation and mutual definition, in which design is not a production of symbiosis but rather takes place as a subprocess within an encompassing symbiosis, the full implication of which resides neither in human or machine but in the system brought into being between these two constituents. In this way an interface theory describes the relations and events through which a system is produced and by which it operates, and so only secondarily pertains to the entities constituting that system. In a human-machine system the questions of what is human and what is machine are only posed operatively; the interface performs these questions in separation and augmentation. Separation is not a maintenance of existing categories but rather an active defining and setting into relation. Augmentation binds together into a mutual state of being that which was and remains separated. Every communication, transaction, or conveyance that takes place in the activation of the interface is first given force and expression from a constituent entity defined by separation, to be transposed across the interface in a form accessible to a corresponding entity also defined by separation. Such reciprocal transpositions become a kind of communication governed by the interface, and from such communication arises the behavior and intelligence that defines the system.

This is to say that from the separation and augmentation of the interface emerges the system. If one were then to track back through the system to interrogate its constituent entities, one might look from augmentation back to separation to find a defining of those entities. In this way the human-machine interface operatively defines both human and machine, at whatever scale or context; here it works to identify, test, and develop the potential of its constituent entities. As human-machine systems increasingly condition contemporary life, notions of human intelligence, and of machine intelligence as well, increasingly reflect those system-level intelligences that draw upon the human and the machine in their formation, and that produce in their operation the standards of measure by which they are defined as essentially different and yet compatible to one another. And if human-machine systems of any kind are given to possess an evolutionary form of development, one that encompasses within its search space all of the relevant contextualities, materialities, and methodologies that may play in that evolution, one may also find within that search space those qualities of intelligence and life-likeness that now operate in the mutual defining of both human and machine.

System

A system is axiomatically multiscalar; it is at the same time an overall behavior and a relation among constituent parts. Human factors and ergonomics sought to encompass a multidisciplinary response to problems in the systematization of the human-machine relationship. Among its most influential areas of concern were wartime aviation and the pilot-plane system. During World War I these problems included pilot training and selection, the development of aptitude tests (some of which were arguably flight simulators), and the effects of high altitude on cognitive performance. Citing such wartime work, Knight Dunlap, who organized the Psychology Section of the US Army Air Service Medical Research Laboratory while on leave from Johns Hopkins University, described the need for a multidisciplinary approach in response to "the complicated psychological, physiological, and physical problems involved in flying." In a letter to the US National Research Council, Dunlap recommended that subsequent "research on flying personnel, and the development and application of methods of improving flying conditions from the point of view of personnel should be under the direction of a board of scientists, including a physicist, a physiologist, a psychologist, and a competent medical man, with other specialists."[36] Flight would demand of the pilot a wide array of cognitive and perceptual abilities, psychological and physiological responses, and learned and innate behaviors both consciously and tacitly performed. These in turn would be matched against variables in machine performance, instrumentation design, and the demands of wartime aviation. If experimental psychology immediately before World War I could be characterized as culminating in behaviorism—Dunlap was a colleague of John B. Watson—the human-machine system elicited a refocusing of this research trajectory. Instead of making ontological claims over human nature, the means and methods of behaviorist research could instead be turned toward the calibration of human-machine systems.

By the end of World War II the human-machine system was explicitly a subject of study. Psychologist Franklin V. Taylor, who, following World War II, was working on early-warning radar systems as director of the US Naval Research Laboratory, would argue that human-machine systems were best evaluated on their own terms, rather than treating human or machine as separate problems. For Taylor the human operator is best seen as an "organic data transmission and processing link between the mechanical or electronic displays and controls of the machine."[37] While both human and machine would serve as subjects of testing and study, such research would culminate in neither of these separately, but rather in the performance of the human-machine system. Likewise, the system only functions through a continuous testing and measuring of its constituent components. In this way a human-machine system describes a concurrent and mutual measuring of human by machine and of machine by human. Such measurements are only secondarily directed toward establishing a normalization of component parts; rather, they are primarily directed toward a calibration of components within a system-level performance. As such, the aim of human-machine systems research could be described as seeking out discrepancies and spaces of reconciliation in the capability and behavior of human and machine.

One example of this in postwar systems research is the Kinalog Display System proposed by the General Dynamics Corporation in 1959.[38] Like the attitude indicator it was meant to replace within a suite of cockpit instrumentation, the Kinalog would convey to the pilot the orientation or attitude of the aircraft with respect to the earth, essential information when visibility is impaired in flight. Also like the attitude indicator, the Kinalog would represent the relation of the wings of the aircraft to an artificial horizon (an early name for the attitude indicator) in order to indicate pitch and roll. Yet where standard attitude indicators display fixed

Right turn at 45° bank angle.

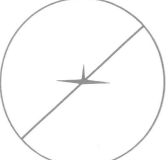

Fig 1. Inside-out attitude display.

Just after reaching the 45° bank angle in coordinated right turn.

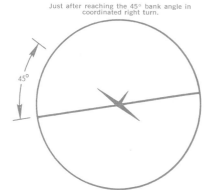

Fig 4. Kinalog attitude display.

Right turn at 45° bank angle.

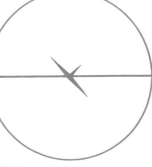

Fig 2. Outside-in attitude display.

Ten seconds after entry into steady 45° bank coordinated turn to the right.

Fig 5. Kinalog attitude display.

During initiation of a 45° bank to the right for a coordinated turn.

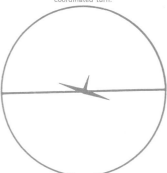

Fig 3. Kinalog attitude display.

Thirty seconds after entry into steady 45° bank coordinated turn to the right.

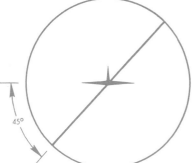

Fig 6. Kinalog attitude display.

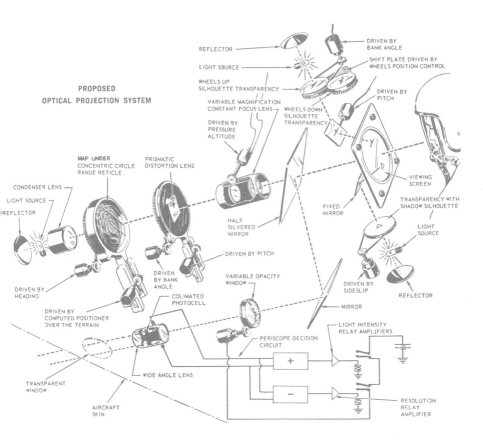

PROPOSED
OPTICAL PROJECTION SYSTEM

REFLECTOR

LIGHT SOURCE

WHEELS UP
SILHOUETTE TRANSPARENCY

VARIABLE MAGNIFICATION
CONSTANT FOCUS LENS

DRIVEN BY
PRESSURE
ALTITUDE

DRIVEN BY
BANK ANGLE

SHIFT PLATE DRIVEN BY
WHEELS POSITION CONTROL

DRIVEN BY
PITCH

WHEELS DOWN
SILHOUETTE
TRANSPARENCY

MAP UNDER
CONCENTRIC CIRCLE
RANGE RETICLE

PRISMATIC
DISTORTION LENS

CONDENSER LENS

LIGHT SOURCE

REFLECTOR

VIEWING
SCREEN

TRANSPARENCY WITH
SHADOW SILHOUETTE

FIXED
MIRROR

HALF
SILVERED
MIRROR

LIGHT
SOURCE

DRIVEN BY PITCH

DRIVEN
BY BANK
ANGLE

VARIABLE OPACITY
WINDOW

DRIVEN BY
SIDESLIP

REFLECTOR

DRIVEN BY
HEADING

COLIMATED
PHOTOCELL

MIRROR

DRIVEN BY
COMPUTED POSITIONER
OVER THE TERRAIN

LIGHT INTENSITY
RELAY AMPLIFIERS

PERISCOPE DECISION
CIRCUIT

+

TRANSPARENT
WINDOW

WIDE ANGLE LENS

−

RESOLUTION
RELAY
AMPLIFIER

AIRCRAFT
SKIN

FIGURE 3.1

—

KINESTHETICS OF A 45-DEGREE BANKED
TURN, 1959. "THE FIRST THING A PILOT
FEELS IS THAT THE AIRPLANE HAS BEEN
TILTED TO THE RIGHT. HIS BODY REC-
OGNIZES THAT THE AIRCRAFT IS TRULY
THE MOVING MEMBER. BY THE TIME HE
HAS REACHED THE DEGREE OF BANK HE
DESIRES, HIS BODY HAS ALREADY STARTED
TO ADAPT, BUT AT A RATE MUCH SLOWER
THAN THE BANKING OF THE AIRCRAFT."

SOURCE: FOGEL, "A NEW CONCEPT: THE KINALOG
DISPLAY SYSTEM," 31, QUOTE ON 30.

FIGURE 3.2

—

MECHANISMS OF OPTICAL PROJECTION,
1959. "'MOST NATURAL' IS AN IMPORTANT
CONCEPT. IT IMPLIES THE METHOD OF
INTEGRATION WHICH THE PILOT HAS LIVED
WITH ALL OF HIS LIFE."

SOURCE: FOGEL, "A NEW CONCEPT: THE KINALOG
DISPLAY SYSTEM," 34, QUOTE ON 33.

wings relative to a moving horizon (a display called "inside-out"), with some Russian-made versions displaying a fixed horizon relative to moving wings (called "outside-in"), the Kinalog would render both wings and horizon mobile according to the electromagnetic measurement of g-force in the aircraft. With respect to the pilot-plane system, the Kinalog opened up within the attitude indicator a space in which to address vertigo or spatial disorientation. It sought out discrepancies between the pilot's subjective experience of acceleration in flight and the actual orientation and trajectory of the aircraft; it was specifically designed to "resolve sensory conflict between visual and kinesthetic inputs."[39] While a visual display based on a simple diagrammatic relationship between wing and horizon (this basic diagram remains, even where the Kinalog proposes integrating the horizon with an optically projected microfilm map of the terrain below), the Kinalog would be calibrated to the time of kinesthetic adjustment as the pilot's body reacts to changes in acceleration in the course of maneuvering the aircraft.

Where the standard attitude indicator displays a kind of figure-ground relationship, with one fixed and one mobile element (and one kind of disorientation involves a perceptual reversal of which is fixed and which is mobile), the Kinalog performs a calibrated blurring of figure and ground as the pilot's sensations come into adjustment with the cockpit measurements. While the changes it proposes to the visible cockpit instrumentation panel are as simple as the way two lines relate to one another in a single piece of instrumentation, the Kinalog represented one step toward "a maximum of compatibility between the pilot and cockpit." Designed for the speed of decision making needed for combat aircraft, and looking forward toward the "threshold of entry into space," the Kinalog proposed a strategic compression and integration of flight information such that the pilot "can act almost intuitively with the confidence that his reflex responses will prove to be correct."[40]

As systems proliferate in ever more sophisticated forms, so do the metrics by which both human capabilities and machine performance are evaluated. This testing by the system is neither provisional nor arbitrary; rather, it occurs within the operation of the system. Its measurement is grounded within the ontology of the system itself. Yet in being so, it is also ultimately inaccessible outside the system, insofar as the testing done by the system occurs within its internal functioning. The site of this testing is the interface, as the demarcation that draws the system into a uniform condition. An additional move is then needed to open up the system for external measurement or for measurement of its individual constituents. This external form of measurement is also performed across an interface, albeit one that is externally imposed upon the system as a product of artifice or design. The two forms of testing, internal and external to a system, correspond with these two aspects of the interface, internally defining and externally imposed. With respect to the interface that internally defines the system, the system is ultimately irreducible to the separate activities of its constituent parts, even as these parts are continuously tested in the very existence of the system.

Within a pilot-plane system, the relation of human and machine describes a continuous testing whose mediation takes place across the instrumentation and controls of the cockpit environment. The cockpit as interface renders pilot and plane mutually intelligible and serves as a zone of contact extending into both the machinery and control surfaces of the aircraft and the capabilities of the pilot—including consciously as well as tacitly performed actions, and trained as well as innate sensations and responses. Yet these internal measurings, implicit within system performance, remain internal to that system until opened up to an external expression. They are otherwise unavailable to external representation. Only through an external imposition may the system be opened up to transform an internal interface into an external one or into a surface, even if only for a moment of measuring. It is with the transformation of internal interface into external surface that the activities of one or another of the system's constituents may be separated out and brought to light.

Here the cockpit as interface possesses multiple aspects. Within the human-machine system, the cockpit demarks the singular site of an event, which comprises all those activities that contribute to controlled flight. In this aspect, the event as it occurs across the cockpit is irreducible and unavailable for representation outside the trajectory of flight. Yet the cockpit is also an artifact, an environment produced by design; it is informed and overlaid by multiple lineages of measuring, testing, and data gathering, as well as of iterative and evolving design and production processes with respect to the aircraft, and selection and training regimes with respect to the pilot. In this aspect the cockpit is opened up along multiple points of contact as a device for producing measurements and generating data. This opening up applies as well to the flight simulator (even as the pilot-simulator system may also be viewed as possessing its own interiority). That the cockpit, like the interface in general, spans both of these forms of relation, internally emergent and externally imposed, marks it as a site from which a *system intelligence* is produced. Such system intelligence describes the bridge by which the various intelligences of human and machine may be identified, drawn out, and brought to comparison.

As much as any system possesses in itself a kind of ontology, a coming into being as an event through the internal processes of which each of its constituent parts is defined, exploited, and granted meaning, it is at the same time given as an opening up to measurement and description from the outside. In this way the sciences of complex processes describe *system* and *interface* as pertaining both to a natural process and to the means by which that process may become subject to external description and control. Here science proceeds through iteration toward an ever tighter coupling of simulation and natural process. Likewise, the internally emergent operations of a given human-machine system, while being in themselves embodied within that system and so inaccessible to a complete description from outside, nonetheless draw upon and develop for exploitation new potentialities otherwise latent or hidden within that system's constituent entities, whether human or machine. Insofar as externally imposed operations upon a system track or mimic as intimately as possible its internally emergent operations, they become methods of discovering and exploiting new characteristics and capabilities. As such, the human-machine interface constitutes a means of revealing and drawing forth from either human or machine those qualities, capabilities, and intelligences that otherwise would have remained tacit, dormant, unobserved, or otherwise unavailable to any systemization.

The contemporary proliferation of interfaces, whether by military-industrial projection or market preference, in this way constitutes an unprecedented regime of testing and development. This remains the case even if such interfaces are primarily directed toward producing system behaviors and not component measurements. For the system constitutes a kind of body and embodiment, albeit one already predisposed to being opened up from the inside along the fault line of the interface. This fault line is also the singularity of the system. To paraphrase Deleuze and Guattari, one might then propose a *systemic phylum*, a seeking of singularities across systems and from within its constituent elements of human and machine; following the *machinic phylum*, this would be "a constellation of singularities, prolongable by certain operations, which converge, and make operations converge, upon one or several assignable traits of expression."[41]

Philosophies of embodiment have become an instrumental discourse, speaking not only to a human condition but also to the production and behavior of systems. In *Understanding Computers and Cognition* (1986), computer scientists Terry Winograd and Fernando Flores instrumentalize Heideggerian concepts—e.g., *thrownness, breaking down, readiness-to-hand, presence-to-hand*—as design methodologies in computer science and HCI. They describe "how shifting from a rationalistic to a Heideggerian perspective can radically alter our conception of computers and our approach to computer design."[42] Here the controls of an automobile and their tacit, subconscious use in driving become a model for computer interface design. Citing the recently designed graphical user interfaces of the Xerox Star (1981) and Apple Macintosh (1984) personal computers, Winograd and Flores argue that "within the domains they encompass—text and graphics manipulation—the user is 'driving,' not 'commanding.'"[43] Control becomes tacit and readiness-to-hand becomes a design outcome, a "transparency of interaction" that nonetheless "is not best achieved by attempting to mimic human faculties. In driving a car, the control interaction is normally transparent. You do not think 'How far should I turn the steering wheel to go around that curve?' In fact, you are not even aware (unless something intrudes) of using a steering wheel. Phenomenologically, you are driving down the road, not using controls."[44] Here interface design takes as a problem not only computation, control systems, and the like, but also tacit knowing and readiness-to-hand. For Winograd and Flores "there is a network of equipment that includes my arms and hands, a keyboard, and many complex devices that mediate between it and a screen. None of this equipment is present for me except when there is a breaking down."[45] The interface here is located in human cognition and perception as much as in any device.

Positioning

Computational devices now evolve according to the growing sophistication of their means of charting and harnessing the flows of human attention and intuition. The interface as a cultural form marshals the deployment of computational power into all aspects of human life. The movement from mainframe to personal computer to portable devices is in this way more than the liberation of a processing power once confined to large-scale state or corporate interests—as for example implied in Ridley Scott's "1984" advertisement for the Macintosh. It is a proliferation of code, network, and interface into all aspects of human life. Following the etymology of Foucault's term *dispositif*, every mediated relation given in this proliferation possesses its own specific *positing*, or setting of elements into relation and context, as well as its own specific *positioning*, or performance of that relation as an event.

Among many suitable examples could be cited the computer mouse, described in Engelbart's 1967 patent application as "an X-Y position indicator control for movement by the hand over any surface to move a cursor over the display on a cathode ray tube."[46] While the patent primarily addresses technical specifications, it also suggests the outlines of the human-machine system—or alternatively, its conditions of augmentation. This system requires not only the coupling of hand and mouse, and so the translation of a limited range of movement into machine-readable information, but also the coupling of eye and display, such that the system encompasses as well the couplings of mouse and display and of hand and eye. The computer mouse, then, is designed as means for the "alteration of the display by the human operator in order to deliver instructions to the computer," and more specifically "for accurately indicating the exact position on the visual display so he can make alterations."[47]

The computer mouse was designed as an improvement over the light pen, which in turn had been developed at MIT in 1950 for use with the Whirlwind computer (1945–1952). Under the direction of engineer and dynamic systems theorist Jay Forrester, Whirlwind became the prototype for the first networked, real-time interactive computer system, the SAGE (Semiautomatic Ground Environment) Air Defense System. The project was first funded by the US Navy to be an aircraft simulator; in the words of the light pen's lead designer, Robert Everett, it was to have included "both a cockpit and a very large computer," that could "actually solve the equations of motion and aerodynamics of an aircraft," such that "putting wind tunnel data into the trainer would cause it to fly like an airplane not yet built." The first stage in the development of Whirlwind was driven by "the speed of operation, the amount of equipment, and the dynamic range needed to solve this aircraft stability problem."[48] When the US Navy limited funding in 1950, Whirlwind had developed to the point where it attracted the attention of the Air Force in their search for an air defense system.

"One of the things I think we did first," writes Everett, who later went on to become president of the MITRE Corporation, which took charge of SAGE after its official separation from MIT, "was to connect a visual display to a computer." The display was configured to allow "the machine to select and point the cathode ray tube beam toward any x-y position, after which an intensification pulse would cause a spot to appear on the scope in the place determined by the computer."[49] The scope would display radar data in dots and characters, such as *T* for target aircraft and *F* for fighter; a device was then required for targeting the display. When a joystick-type solution was judged too inaccurate and slow, the concept of the light pen emerged; it was then termed a "light gun" for its primary role in "manual target acquisition."[50] This gun was actually a receiver, held by its operator over the display. With a photocell triggered by a flash on the display, the light gun would transmit this back to the computer as information, and the computer would respond according to its current control settings. In a 1950 summary report, the light gun was described as "a photoelectric device which is placed over the desired spot on a light scope. The next intensification of the selected spot produces a pulse in the light gun, and this pulse is fed into the computer to select the subprogram."[51] In its application in SAGE the subprogram could be, for example, the computation of an interception trajectory.

The light gun would later be redesigned as a "light pen" with help from a team of psychologists led by Licklider, at the time associated with MIT's Project Lincoln (soon after Lincoln Laboratory), into which Whirlwind was incorporated in 1951. Yet in its transition from gun into pen, the light pen retained its function as a targeting device. One member of Licklider's team at the time, psychologist and later Columbia University president William McGill, would describe its use: "Put it over a target blip, press a button on the pen, and acquire the location of the target in the computer." For McGill, Licklider was "always immensely proud of the light pen" and of the role psychology, and in particular notions of human-computer symbiosis, played in its design.[52] The light pen would not only target; it would also facilitate a kind of communication best represented by the drawn line. In his symbiosis paper Licklider had called for an "effective, immediate man-machine communication"; yet, he argued, "the department of data processing that seems least advanced, insofar as the requirements of man-computer symbiosis are concerned, is the one that deals with input and output equipment or, as it is seen from the human operator's point of view, displays and controls."[53] The light pen, then, would be a step in communication with machines that sought "the flexibility and convenience of the pencil and doodle pad or the chalk and blackboard used by men in technical discussion"; it would facilitate a particular relation with the computer, one defined by drawing and drafting as well as handwriting, such that the operator of a computer could "in general interact with it very much as he would with another engineer, except this 'other engineer' would be a precise draftsman, a lightning calculator, a mnemonic wizard, and many other valuable partners all in one."[54]

Licklider looked toward computer-aided design (CAD) as a centrally important step in the development of human-machine symbiosis. He was enthusiastic about Sketchpad, a program designed by Ivan Sutherland for a 1963 MIT doctoral thesis (titled "Sketchpad: A Man-Machine Graphical Communication System"), under the guidance of Claude Shannon. Licklider described a presentation of Sketchpad by Sutherland at a 1962 conference session on "man-computer communication" chaired by Engelbart, as including "the most dramatic on-line graphical composi- tions any of them had seen."[55] Like Licklider, Sutherland looked toward the act of drawing or its augmentation less as an end in itself than as a means of communication with machines; thus Sutherland introduced Sketchpad as a system that "makes it possible for a man and a computer to converse rapidly through the medium of line drawings."[56] Designed specifically for use on one of the Lincoln Laboratories computers (the Lincoln TX2) with the light pen as its primary interface device, Sketchpad either introduced or first integrated a number of the core features now standard in both CAD and the graphical user interface (and so directly influencing projects from Alan Kay's Dynabook to the Xerox Star and the Apple Macintosh), from the use of the light pen either to select visible objects on the screen for subsequent modification or to generate or position objects by dragging the light pen across the display, to the programmed performance of operations such as drawing lines between selected points, setting one line parallel to another, or smoothing curves drawn manually by the light pen. Engelbart's x-y position indicator circumvented a disadvantage of the light pen: where the light pen required the imposition of the hand and device between the eye and display, the mouse would essentially clear the channel between eye and display. In doing so, an overall system of hand/device and eye/display would be moved toward a further integration, to the point where the hand and device coupling becomes internally integrated or tacit within the eye and display coupling, and vice versa.

As an axiom, the interface operates not only as an overcoming or bypassing of resistances of various kinds but also as a seeking out of resistances, as though they were markers of territories to be colonized. It is through such resistances that the differences by which the interface operates are generated. Resistance, whether in human or machine, constitutes the substance upon which a *positioning* may be registered. Engelbart's mouse is in this sense a collection of resistances: from the mechanical, as in the frictional movement of an "idler ball bearing" across a surface and the translation of that movement into x-y components through contact with two wheels "mounted with their axis perpendicular to each other"; to the electrical, as with the potentiometers to which each of these wheels is attached, where resistance performs the transduction of the wheels' mechanical movement into electric signals; to the ergonomic, where a kind of resistance may be found to operate insofar as the mouse constituted an utterly new form of tool device, one never before encountered, at once contoured to be welcoming to use through its being "adapted to be held in the hand," and at the same time foreign in eliciting from the hand a novel and specific form of gestural and articulated movement.[57] Other such plays upon resistances—circumventions and enhancements, overcomings and amplifications—may also be found within the hand as it relates to the device, in those aspects of the hand the device draws upon and exposes for use within the overall human-machine system. Here one may consider any aspect of the hand as it relates or may relate to the device: from its skeletal, muscular, and nervous system physiologies, to the senses it represents, tactile and kinesthetic, and to the relations of these to the eye and visual sensation; and especially the interrelations of all of these in the training of the hand with respect to the device.

A different form of training, or entrainment, may then be found with each device as it in this way becomes ready-to-hand. While Engelbart's patent addresses the technological specifications of his "x-y position indicator," one might also imagine the corresponding and yet-to-be-written specifications addressing the training of the human operator. And within such entrainment may also be found a lineage, for example from light gun to light pen to mouse, just as the referent upon which each of these is based passes from a targeting device to a drawing device to another less referential form of positioning, one that is at once more abstract with respect to earlier models of human behavior (drawing; aiming and firing) and more concrete with respect what may be termed its system-level embodiment. That is, the evolution of the computer mouse as a device may be found to have followed to some critical extent those qualities, behaviors, and logics drawn from a still-developing intermediary zone between human and machine, so as to register both a level of machine development and a level of human training, in both resistances and possibilities, even as it carries with it a lineage of the contexts and capabilities from which it was developed, such that within the use of the mouse also exist traces of the drawn line with pen in hand and the aim of a gun and the pulling of a trigger.

In this constellation of decision making rendered in gestures, where communicating is as drawing is as targeting, one might add the calculated lines of intercept given in air defense systems, and perhaps finally the originary site of the cockpit, with its flight controls evolving into the joystick of a training simulator. In each of these cases, what could be described as intelligence, as a capacity for differentiable communication between human and machine with respect to a given system of interface, is on the human side of the equation constituted to a large extent of capabilities that are only tacitly known and learned: not as a result or product of a given process of rationalization, but from an expansive process of seeking along the turbulences and flows of an evolutionary lineage whose substance is both human and machine. It is here that the emerging figure of a *genius of augmentation* may be found, revealed in its various aspects, both tacitly and consciously known, originating both from within and without the contemporary subject, to which the technological appears both as genial and as an internalized self-alienation. For within augmentation is also a separation, and the genius emergent from within this encountering of human and machine may be found to both preside over those obscurances of humanness in the advent of its technological augmentation—and so over that particular fog in which the purveyors of technological positivities (artificial intelligences, technological singularities, human-machine symbioses, and so on) symptomatically lose sight of that which is human in its relation to the machine—and to at the same time foreground humanness as a problematic. Here augmentation reveals as-yet-unrevealed aspects of humanness that constitute the fluid boundary of that which may and that which may not be brought into communication with the machine. This is even more so today, where the current line of technological development by which humanness encounters its augmented genius aims toward ambience and ubiquity, so heralding the advent of sustained, expanded, and ever more finely elaborated operations for the concurrent elisions and differentiations of human and machine.

Following Heidegger, augmentation is a *bringing forth* into existence of that which otherwise would not have existed; as a kind of *poiesis*—citing Plato that "all creation or passage of nonbeing into being is poetry or making" [58]—processes of augmentation may be viewed as not only bringing into being the augmented condition itself, but also at the same time producing or revealing, as a kind of presencing, those aspects of the elements brought together in augmentation through which augmentation occurs, whether in ego or genius, human or machine. What is revealed is that which is embodied, singular, or exceptional within each of these elements, as it is according to these qualities that augmentation is both resisted and allowed to come into being. In this way the contemporary arc of technological development, regarding the relation between human beings and machines as focused upon the figure of the interface, may be described as a seeking out of exceptions, which may then serve both as the limits of possible development and as the raw material to be drawn into further development and elaboration.

Here one may point to another defining of the poetic; in reference to "the dialectical battle through which the exception emerges from the universal," Søren Kierkegaard proposes that "a poet is such an exception" as well as being a "transition" to the exception.[59] Schmitt would draw upon Kierkegaard's exception in his theory of the political, in which sovereignty is defined according to the ability to determine the exception; likewise,

Agamben describes the "paradox of sovereignty": that the "sovereign is, at the same time, outside and inside the juridical order."[60] Kierkegaard writes of the exception that it "explains the universal and itself, and when one wants to really study the universal, one need only examine a legitimate exception ... if one cannot explain them, then neither can one explain the universal. One generally fails to notice this, because one does not usually grasp the universal passionately, but only superficially. The exception, on the other hand, grasps the universal with intense passion."[61] For Schmitt and Agamben, what is defined by the exception is both the limit and the fulfillment of power and of the attribution of sovereignty. Yet what emerges as paradoxical in this defining is not only the relation of the exception to the universal, which the exception at the same time separates itself from and defines, but also the extent to which a form of ambiguity coalesces around the exception, an ambiguity that is always found in discourses of power between the attribution of an identity to a thing and the emergence of a thing through an event. Along with its designation from without, as by the sovereign, the exception also possesses a countervailing emergence from within, as with the poetic or *poiesis*. In this sense the exception may refer to that form of productive creation which takes place within the ambiguity of the liminal condition or the threshold. It is in this ambiguity that the exception holds to the passage between nonbeing and being. Kierkegaard writes: "My poet finds justification in that existence absolves him in that instant when he wishes, in a sense, to destroy himself. His soul then wins a religious resonance."[62] A similar movement might then also be read across the interface, whose subject must first endure a separation that is also a fragmentation, before achieving, at least in a spectral, momentary sense, a completeness in augmentation. In the relation of human beings to machines, it is the potential value of an interface theory to trace out such relations in all of the flows and turbulences in which they occur, and to find in them the linkages to systems of power that might otherwise remain concealed.

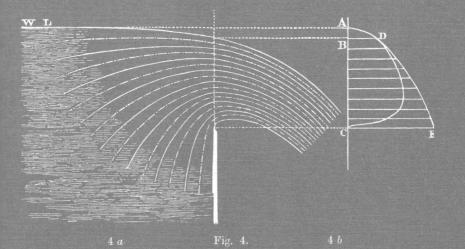

W L   A

D

B

C   E

4 a   Fig. 4.   4 b

NOTES

1 THE SUBJECT OF THE INTERFACE

1. This definition of human-computer interface (HCI) is found in *A Dictionary of Computing*, 6th ed. (Oxford: Oxford University Press, 2008), 240. The definition provides as synonyms human-system interface (HSI), human-machine interface (HMI), and man-machine interface (MMI).

2. For example, interaction design is defined by Gerhard M. Buurman as "the paradigmatic representation, aesthetic/sensory presentation, and technological implementation of computer-supported perceptions (cognition and sensorics) and actions (actuator systems and visualization) in mixed media." Computer scientist Paul Dourish describes how "interactional approaches conceptualize computation as an interplay between different components, rather than the fixed and prespecified paths that a single, monolithic computational engine might follow." Gerhard M. Buurman, "Introduction," in *Total Interaction: Theory and Practice of a New Paradigm for the Design Disciplines*, ed. Gerhard M. Buurman (Boston: Birkhäuser, 2005), 9; Paul Dourish, *Where the Action Is: The Foundations of Embodied Interaction* (Cambridge: MIT Press, 2001), 4.

3. The problem of agency posed by interactive technology may be found in Donna Haraway's cyborg, "the figure born of the interface of the automaton and autonomy." Likewise, anthropologist of technology Lucy A. Suchman describes how an extensive body of work in the social sciences on interactivity is "frequently summarized by the proposition that humans and artifacts are *mutually constituted*." Donna Haraway, *Primate Visions: Gender, Race, and Nature in the World of Modern Science* (New York: Routledge, 1989), 139; Lucy A. Suchman, *Human-Machine Reconfigurations: Plans and Situated Actions*, 2nd ed. (Cambridge: Cambridge University Press, 2007), 268. Italics are in the original.

4. Ovid, *Fasti*, trans. James George Frazer (New York: G. P. Putnam's Sons, 1931), 11 (subsequent page references in the text are to this edition). The Latin term *fasti*, which through its root *fas* signifies divine law, gives a sense of the Roman calendar as a regulator of legal power, since the calendar determined the *fasti dies*, or lawful days, on which legal business could be transacted. For example, the narrator of Ovid's *Fasti* asks Janus why the first day of the year is "not exempt from lawsuits," to which Janus replies: "I assigned the birthday of the year to business, lest from the auspice idleness infects the whole." Ovid, *Fasti*, 15. See also *A Dictionary of Greek and Roman Antiquities*, ed. William Smith and Francis Warre Cornish (London: John Murray, 1898), 521–522.

5. Georges Dumézil, *Archaic Roman Religion*, vol. 1, trans. Philip Krapp (1966; Baltimore: Johns Hopkins University Press, 1996), 328.

6. Ibid, 332. For Ovid the doors were shut to restrain war and opened to bring peace: "When I choose to send forth peace from tranquil halls, she freely walks

157

the ways unhindered. But with blood and slaughter the whole world would welter, did not the bars unbending hold the barricaded wars." Ovid, *Fasti*, 11.

7. Dora and Erwin Panofsky, *Pandora's Box: The Changing Aspects of a Mythical Symbol* (1956; Princeton: Princeton University Press, 1978), 3. For the Panofskys, Pandora was needed "to corroborate the doctrine of original sin by a classical parable" (11) and so changed to reflect theological arguments, while for E. H. Gombrich, Pandora represented "the continued creativity of mythological invention after the official demise of paganism" (see Gombrich's review of the Panofskys' book in *Burlington Magazine* 99, no. 653 [August 1957]: 280).

8. James George Frazer, *Lectures on the Early History of the Kingship* (London: Macmillan, 1905), 285.

9. Ibid, 286.

10. Ibid, 288.

11. Arthur Bernard Cook, "Zeus, Jupiter, and the Oak," *Classical Review* 18, no. 7 (October 1904): 367.

12. Ibid, 368.

13. Dumézil, *Archaic Roman Religion*, 100.

14. Ibid, 328.

15. Ibid, 327. Another version of Augustine's quotation of Varro reads: "The reason is that the start of things rests with Janus but the fulfillment with Jupiter, who is therefore rightly held to be the sovereign power. For the fulfillment surpasses the beginning; the beginning has precedence in time, but the fulfillment is superior in dignity." Augustine, *City of God*, trans. Henry Bettenson (London: Penguin Books, 2003), 265–266.

16. Dumézil, *Archaic Roman Religion*, 327.

17. Edwin G. Boring, "The Nature and History of Experimental Control," *American Journal of Psychology* 67, no. 4 (December 1954): 589. The first control experiment was described as early as 1648 with Pascal's use of the recently invented barometer to measure the pressure of the air. The first systematic treatment of the control experiment was by John Stuart Mill in 1843 in his description of the "method of difference," of one of four logical methods for use in experimentation. In this method, control operates between two observations to hold all conditions and trajectories constant while maintaining one critical difference. Mill would write of this method: "It thus appears that in the study of various kinds of phenomena which we can, by our voluntary agency, modify or control, we can in general satisfy the requisitions of the Method of Difference; but that by the spontaneous operations of nature those requisitions are seldom fulfilled." John Stuart Mill, *A System of Logic, Ratiocinative and Inductive* (1843; New York: Harper & Brothers, 1874), 282.

18. The philosopher Giorgio Agamben describes the apparatus as "a decisive term in the strategy of Foucault's thought," while Deleuze relates how "Foucault's

philosophy is often presented as an analysis of concrete social apparatuses [*dispositifs*]." Giorgio Agamben, *What Is an Apparatus? and Other Essays*, trans. David Kishik and Stefan Pedatella (Stanford: Stanford University Press, 2009), 2; Gilles Deleuze, "What Is a *dispositif*?," in Timothy J. Armstrong, ed. and trans., *Michel Foucault: Philosopher* (New York: Routledge, 1992), 159.

19. Michel Foucault, *Power/Knowledge: Selected Interviews and Other Writings, 1972–1977*, ed. Colin Gordon (New York: Pantheon Books, 1980), 194, 195.

20. Foucault describes a positivity [*positivité*] as "an incomplete, fragmented figure" or "the dispersion of an exteriority" or "the specific forms of an accumulation" that characterize the concrete mechanisms of power in processes of subjectification. Michel Foucault, *The Archaeology of Knowledge and The Discourse on Language*, trans. A. M. Sheridan Smith (New York: Pantheon Books, 1972), 125.

159

21. Agamben, *What Is an Apparatus?*, 13, 14.

22. Giorgio Agamben, *Homo Sacer: Sovereign Power and Bare Life*, trans. Daniel Heller-Roazen (Stanford: Stanford University Press, 1998), 5.

23. Agamben, *What Is an Apparatus?*, 19–23.

24. "All significant concepts of the modern theory of the state are secularized theological concepts not only because of their historical development—in which they were transferred from theology to the theory of the state, whereby, for example, the omnipotent God became the omnipotent lawgiver—but also because of their systematic structure, the recognition of which is necessary for a sociological consideration of these concepts." Carl Schmitt, *Political Theology: Four Chapters on the Concept of Sovereignty*, trans. George Schwab (Chicago: University of Chicago Press, 1985), 36.

25. Agamben traces two conceptual and etymological lines from apparatus to theological notions of power, as part of what he calls a "theological genealogy of economy." First, *dispositif* and the earlier *positivité* are traced to Hegel's distinction between natural and positive religion, the former denoting the "immediate and general" relation to the divine available to reason, and the latter the culturally and historically situated beliefs and rituals involved in command and obedience, "externally imposed on individuals." Second, *dispositif* is traced to the Greek *oikonomia* (economy) via its Latin translation *dispositio*, to denote a theological separation between being and acting, in which "economy" referred to the interventions of the church in the management or disposition of worldly events. Agamben, *What Is an Apparatus?*, 4, 8.

26. Schmitt, *Political Theology*, 5, 13.

27. Carl Schmitt, *The Concept of the Political*, trans. George Schwab (Chicago: University of Chicago Press, 1996), 26, 27.

28. See Agamben, *Homo Sacer*, 71–74.

29. Ibid., 65; see also Walter Benjamin, "Critique of Violence" (1921), in *Reflections: Essays, Aphorisms, Autobiographical Writings* (New York: Harcourt Brace Jovanovich, 1978), 299.

30. Schmitt, *Political Theology*, 15.

31. Ibid., 5.

32. Hannah Arendt, *The Origins of Totalitarianism* (1951; New York: Harcourt, 1976), 245.

33. Agamben, *What Is an Apparatus?*, 19.

34. Johan Huizinga, *Homo Ludens: A Study of the Play Element in Culture* (Boston: Beacon Press, 1955), 10.

35. Ibid., 8.

36. Roger Caillois, *Man, Play and Games*, trans. Meyer Barash (Urbana: University of Illinois Press, 2001), 9–10.

37. Ibid., 83.

38. Huizinga, *Homo Ludens*, 173.

39. Caillois, *Man, Play and Games*, 36.

40. Ibid., 8.

41. Huizinga, *Homo Ludens*, 8.

42. Caillois, *Man, Play and Games*, 163.

43. Katie Salen and Eric Zimmerman, *Rules of Play: Game Design Fundamentals* (Cambridge: MIT Press, 2004), 95.

44. Jesper Juul, *Half-Real: Video Games between Real Rules and Fictional Worlds* (Cambridge: MIT Press, 2005), 164, 167, 163. Italics are in the original.

45. For Galloway this classification system is not set of rules, but rather a provisional division into "four moments, four suggestions" that address the range of actions that constitute game play. A "diegetic machine act" includes the production of ambience and cinematic moments to transmit information or set atmosphere; a "nondiegetic operator act" includes configuration or simulation, where the operator most directly addresses game algorithms; a "diegetic operator act" includes the actions of the operator within the game world; and a "nondiegetic machine act" ("the most interesting category") includes those elements outside the game world necessary for its creation, such as the entrance and exit into the game, and provides the most direct experience of the game as coded. See the table found in Alexander R. Galloway, *Gaming: Essays on Algorithmic Culture* (Minneapolis: University of Minnesota Press, 2006), 38.

46. Clifford Geertz, *The Interpretation of Cultures* (New York: Basic Books, 1973), 10.

47. This class of games might more accurately be described as *interface games*. Where the term *video games* privileges the visual and display system, and *computer* or *console games* the gaming machine, *interface games* identifies the human-machine relation as a game space, and emphasizes the passage of all gaming acts across an interface.

48. McKenzie Wark, *Gamer Theory* (Cambridge: Harvard University Press, 2007), 19, 18.

49 Franz Reuleaux, *The Kinematics of Machinery: Outlines of a Theory of Machines*, trans. Alex B. W. Kennedy (London: Macmillan, 1876), 35, 40. Lewis Mumford cites this definition of machine in *Technics and Civilization* (1934; New York: Harcourt Brace, 1963), 9.

50. Mumford, *Technics and Civilization*, 9, 10.

51. Ibid., 10.

52. Ibid., 10, 11.

53. Reuleaux, *The Kinematics of Machinery*, 40.

54. Ibid., 40, 41.

55. Martin Heidegger, "The Question Concerning Technology," in *Basic Writings* (New York: Harper & Row, 1977), 302. This passage is cited by Agamben in relating enframing (*Gestell*) to apparatus (*dispositif*) and positivity (*positivité*). Agamben, *What Is an Apparatus?*, 12.

56. Gilles Deleuze and Félix Guattari, *A Thousand Plateaus: Capitalism and Schizophrenia*, trans. Brian Massumi (1980; Minneapolis: University of Minnesota Press, 1993), 406. Italics are in the original.

57. Ibid., 409, 406, 409.

58. Ibid., 409. Italics in the original.

59. Manuel De Landa, *War in the Age of Intelligent Machines* (New York: Zone Books, 1991), 2–3.

60. Huizinga, *Homo Ludens*, 213, 212.

61. Ibid., 8. If play involves a "*disinterestedness*" with respect to "'ordinary' life" in that "it stands outside the immediate satisfaction of wants and appetites" or "*outside* the purely physiological" as a kind of "intermezzo, an *interlude* in our daily lives," it is only because it performs from within ordinary physiological life a transposition and transfiguration of those wants and appetites in the production of a specific form of activity. Likewise, if play involves a "secludedness" or "limitedness" expressed according to its "locality and duration" with respect to ordinary life, where it is "'played out' within certain limits in time and place," it is only on account that its production of activity "contains its own course and meaning," and springs from those rules that "determine what 'holds' in the temporary world circumscribed by play." Ibid., 9, 11. Italics are in the original.

62. Ibid., 10.

63. Hillel Schwartz, *The Culture of the Copy: Striking Likenesses, Unreasonable Facsimiles* (New York: Zone Books, 1996), 268.

64. Ibid., 260.

65. Huizinga, *Homo Ludens*, 208–211.

66. Leo Strauss, "Notes on *The Concept of the Political*" (1932, trans. 1965), trans. J. Harvey Lomax, reprinted in Schmitt, *The Concept of the Political*, 105. Italics are in the original.

67. Schmitt, *The Concept of the Political*, 64, 65.

68. Huizinga, *Homo Ludens*, 210.

69. Schmitt, *The Concept of the Political*, 66, 67.

70. Ibid., 67.

71. Strauss, "Notes on *The Concept of the Political*," 106.

72. Caillois, *Man, Play and Games*, 14, 17, 23, 19.

73. Roger Caillois, "Mimicry and Legendary Psychasthenia" (1938), trans. John Shepley, *October* 31 (Winter 1984): 23. Italics are in the original.

74. Ibid., 31, 30. Italics are in the original.

75. Ibid., 28. Italics are in the original.

76. Caillois, *Man, Play and Games*, 27, 13, 56, 33, 27.

2 THE FORMING OF THE INTERFACE

1. James Thomson, *Collected Papers in Physics and Engineering*, ed. Joseph Larmor and James Thomson (Cambridge: Cambridge University Press, 1912), 327. This quote is taken from a section dated May 10, 1869, under the heading "Notes and Queries—On Gases, Liquids, Fluids: Unpublished notes bearing on [chemist and physicist Thomas] Andrews' experiments."

2. Ibid., 65. This quote was originally published in 1876 in a paper titled "Improved Investigations on the Flow of Water through Orifices with Objections to the Modes of Treatment Commonly Adopted."

3. Ibid., 424. This quote was originally published in 1877 in a paper titled "On the Jointed Prismatic Structure in Basaltic Rocks."

4. James Clerk Maxwell, *Theory of Heat*, new imprint (London: Longmans, Green, 1902), 95.

5. Cf. Maxwell, *Theory of Heat*, 3rd ed. (London: Longmans, Green, 1872), 95; and 4th ed. (London: Longmans, Green, 1875), 95.

6. James Thomson, "On the Flow of Water in Uniform Regime in Rivers and Other Channels," *Proceedings of the Royal Society of London* 28 (1878–1879): 121.

7. Ibid., 121n.

8. This correspondence between James Clerk Maxwell and William Thomson (Lord Kelvin) may be found in A. T. Fuller, "Maxwell's Glasgow Transcripts: Extracts Relating to Control and Stability," *International Journal of Control* 43, no. 5 (May 1986): 1594. The original source is a letter from James Clerk Maxwell to William Thomson dated November 24, 1857, found as a manuscript at the Glasgow University Library, Kelvin papers, box M10.

9. Ibid. The italics are in Fuller's publication of Maxwell's letter.

10. Navier-Stokes equations describe fluid motion in terms of velocities rather than the positions of particles and are in most cases nonlinear; the only occasions where they may resolve to a mathematically exact solution occurs in those situations where nonlinear terms may be removed and the equations treated as linear. Mathematical exactness here is found only to correspond with flows that are highly viscous and so resistant to turbulence (technically, they have a low Reynolds number; this ratio, between the inertial forces of a fluid and its viscosity or internal resistance, may also be used to predict the onset and scale of turbulence). In all other cases, experiment and approximation are needed.

11. Fuller, "Maxwell's Glasgow Transcripts," 1595.

12. Maxwell, *Theory of Heat*, 3rd ed., 308, 309.

13. Ibid., 308.

14. Ibid., 308–309.

15. William Thomson (Lord Kelvin), "Kinetic Theory of the Dissipation of Energy," *Nature* 9, no. 232 (April 9, 1874): 442n.

16. Iannis Xenakis, *Arts/Sciences: Alloys: The Thesis Defense of Iannis Xenakis*, trans. Sharon E. Kanach (Hillsdale, NY: Pendragon Press, 1985), 71.

17. Herbert Jennings Rose, *Ancient Greek Religion* (London: Hutchinson's University Library, 1948), 109.

18. Momigliano, following Weinstock, writes: "The word 'numen,' originally meaning 'motion,' became a religious term in Rome only in the first century B.C., probably under the influence of the new theology and as a corresponding term to 'daimon.'" Arnaldo Momigliano, "Georges Dumézil and the Trifunctional Approach to Roman Civilization," *History and Theory* 23, no. 3 (October 1984): 319–320.

19. Stephen Weinstock, "Review: H. J. Rose, *Ancient Roman Religion*," *Journal of Roman Studies* 39 (1949): 167.

20. Rose, *Ancient Roman Religion*, 15.

21. Weinstock, "Review," 166.

22. Rose, *Ancient Greek Religion*, 109.

23. Ibid., 110.

24. Ibid.

25. Kelvin, "Kinetic Theory of the Dissipation of Energy," 442.

26. I bid.

27. Maxwell, *Theory of Heat*, 255–257.

28. Kelvin, "Kinetic Theory of the Dissipation of Energy," 442–443.

29. These definitions of daemon are drawn, respectively, from: *Webster's New World Computer Dictionary*, 10th ed. (Indianapolis: Wiley, 2003), 97; *Dictionary of*

Computer and Internet Terms, 10th ed. (Hauppauge, NY: Barron's Educational Series, 2009), 124; A Dictionary of Computing, 6th ed. (Oxford: Oxford University Press, 2008), 123; and Microsoft Computer Dictionary, 5th ed. (Redmond, WA: Microsoft Press, 2002), 140.

30. Alan M. Turing, "Computing Machinery and Intelligence," Mind: A Quarterly Review of Psychology and Philosophy 59, no. 236 (October 1950): 450.

31. Ibid.

32. Ibid., 451.

33. Kelvin, "Kinetic Theory of the Dissipation of Energy," 443.

34. Gaston Bachelard, Water and Dreams: An Essay on the Imagination of Matter, trans. Edith R. Farrell (Dallas: Dallas Institute of Humanities and Culture, 1983), 159–185. In the chapter titled "Violent Water" Bachelard treats the violence of water in psychoanalytic terms, although unfortunately as a masculine phenomenon.

35. bid., 6.

36. Gilles Deleuze and Félix Guattari, A Thousand Plateaus: Capitalism and Schizophrenia, trans. Brian Massumi (1980; Minneapolis: University of Minnesota Press, 1993), 363.

37. Ibid., 554n24. The original quotation is found in Michel Serres, La naissance de la physique dans le texte de Lucrèce: Fleuves et turbulences (Paris: Éditions de Minuit, 1977), 106.

38. Michel Serres, Detachment, trans. Genevieve James and Raymond Federman (1983; Athens: Ohio University Press, 1989), 9.

39. Michel Serres, Genesis, trans. Genevieve James and James Nielson (1982; Ann Arbor: University of Michigan Press, 1995), 98.

40. "La mer est le cycle d'Aphrodite. On comprend ici pourquoi Lucréce l'invoque en commençant: tout vivant émerge de la distribution première, fluide originaire." Michel Serres, La distribution (Hermes IV) (Paris: Éditions de Minuit, 1977), 160.

41. Serres, Genesis, 98.

42. "Les philosophes de l'âge contemporain sont des philosophes du réservoir. De la circulation, de choses stockées, au réservoir." Serres, La distribution, 162.

43. Serres, Genesis, 109.

44. Ibid., 100. See also Michel Serres, The Birth of Physics, trans. Jack Hawkes (1977; Manchester: Clinamen Press, 2000), 27–31.

45. Deleuze and Guattari, A Thousand Plateaus, 474.

46. See Claude E. Shannon, "A Mathematical Theory of Communication," Bell Technical Journal 27 (July and October 1948): 379–423, 623–656.

47. Norbert Wiener, Cybernetics: or Control and Communication in the Animal and the Machine (Cambridge: MIT Press, 1948), 39.

48. Leó Szilárd, "Über die Entropie Verminderung in einem thermodynamischen System bei Eingriffen intelligenter Wesen," *Zeitschrift für Physik* 53, nos. 11–12 (November 1929): 840.

49. Erwin Schrödinger, *What Is Life? with Mind and Matter and Autobiographical Sketches* (Cambridge: Cambridge University Press, 1992), 70, 71.

50. Claude E. Shannon and Warren Weaver, *The Mathematical Theory of Communication* (Urbana: University of Illinois Press, 1949), 12. Italics are in the original. Note that what was in 1948 "*a* mathematical theory of communication" has by 1949 become "*the* mathematical theory of communication."

51. Ibid., 42.

52. Walter Benjamin, "The Work of Art in the Age of Mechanical Reproduction," in *Illuminations: Essays and Reflections*, ed. Hannah Arendt, trans. Harry Zohn (New York: Schocken Books, 1968), 222.

53. Wiener, *Cybernetics*, 11–12. Italics are in the original.

54. Wiener was likely unaware of Ampère's use of *cybernétique* to describe a science of governing. See Otto Mayr, "Maxwell and the Origins of Cybernetics," *Isis* 62, no. 4 (Winter 1971): 425n1.

55. André-Marie Ampère, *Essai sur la philosophie des sciences, ou, Exposition analytique d'une classification naturelle de tout les connaissances humaines* (Paris: Chez Bachelier, 1834), 258n. While the science of politics is discussed in this passage from the 1834 edition, the word *cybernétique* is found only in a large, foldout table at the end of the book, where Ampère's classification system is presented in full; however, a short explanation of the etymology of *cybernétique*, which ties the word both to its Greek root and the word *govern*, is given in the posthumously published 1843 edition.

56. Mayr, "Maxwell and the Origins of Cybernetics," 443.

57. Maxwell's "On Governors" had been published in a prominent journal and included in the 1890 edition of his collected works. For Otto Mayr the obscurity of the paper resulted from its reliance on "mechanisms devoid of industrial significance" that "were neither adequately explained or identified," as though meant to be "understood only by insiders who knew these mechanisms through personal information." A. T. Fuller similarly describes the paper as "allusive and not easily accessible," speculating that Maxwell most likely expected one of his colleagues, most likely Thomson, Kelvin, William Siemens, or Fleeming Jenkin, to "take up and expand on arguments which he had only sketched." See Mayr, "Maxwell and the Origins of Cybernetics," 425–427; and A. T. Fuller, "Maxwell's Treatment of Siemens's Hydraulic Governor," *International Journal of Control* 60, no. 5 (November 1994): 882–883.

58. For Mayr writing in 1971, the identity of the machine Maxwell refers to in "On Governors" as a "water-break" was unknown. It would only be positively identified as Thomson's "centrifugal pump regulator" by Fuller in 1986. See Mayr,

"Maxwell and the Origins of Cybernetics," 427n11; Fuller, "Maxwell's Glasgow Transcripts," 1611.

59. James Clerk Maxwell, "On Governors," *Proceedings of the Royal Society* 16, no. 100 (1868): 270.

60. "Governors" were mentioned in a report dated August 23, 1863, by a committee on electrical standards appointed by British Association for the Advancement of Science. Members of this committee included Maxwell, Kelvin, Balfour Stewart, H. C. Fleeming Jenkin, and Charles William Siemens. See Fuller, "Maxwell's Treatment of Siemens's Hydraulic Governor," 864–865.

61. Maxwell, "On Governors," 270.

62. Mayr, "Maxwell and the Origins of Cybernetics," 427. Italics are in the original.

63. Maxwell's treatment of the *moderator-governor* distinction—which followed earlier distinctions made by Franz Reuleaux (which he termed *static* and *astatic*) and Léon Foucault (termed *nonisochronous* and *isochronous*)—was perhaps the most remarked-upon aspect of "On Governors" at the time it was published. The distinction today would be expressed as between proportional control and proportional-integral control. See Mayr, "Maxwell and the Origins of Cybernetics," 427–428.

64. For example, what Maxwell describes as "governing power" might be equivalent to "open loop gain." Ibid., 443n42.

65. The quote is from Maxwell. The paper was originally presented in January 1868. Mayr, "Maxwell and the Origins of Cybernetics," 426.

66. James Clerk Maxwell, "On Reciprocal Diagrams in Space, and Their Relation to Airy's Function of Stress," in *The Scientific Papers of James Clerk Maxwell*, ed. W. D. Nevin (1890; New York: Dover, 1952), 102–103.

67. James Clerk Maxwell, "On Reciprocal Figures and Diagrams of Force," in *The Scientific Papers of James Clerk Maxwell*, 522–523. This paper was originally published in *Philosophical Magazine*, ser. 4, vol. 27 (1864): 250–261.

68. Arturo Rosenblueth, Norbert Wiener, and Julian Bigelow, "Behavior, Purpose, and Teleology," *Philosophy of Science* 10, no. 1 (January 1943): 18. Italics are in the original.

69. Ibid.

70. See Peter Galison, "The Ontology of the Enemy: Norbert Wiener and the Cybernetic Vision," *Critical Inquiry* 21, no. 1 (Autumn 1994): 228–266.

71. Rosenblueth, Wiener, and Bigelow, "Behavior, Purpose, and Teleology," 19.

72. Ibid., 24.

73. Ibid., 19, 20.

74. Ibid., 22.

75. Ibid., 22, 23.

76. Ibid., 24.

77. Fuller, "Maxwell's Treatment of Siemens's Hydraulic Governor," 863; Thomson, *Collected Papers in Physics and Engineering*, 12–13.

78. Maxwell, "On Governors," 270.

79. Fuller, "Maxwell's Treatment of Siemens's Hydraulic Governor," 882.

80. Fuller argues that Siemens's neglect of Thomson's previous work may have been "the spur which made Maxwell at last put into print his own account of governors." He also points out that Maxwell writes of Thomson having "invented" his water-break where Siemens only "described" his hydraulic brake. See Fuller, "Maxwell's Glasgow Transcripts," 1611; Fuller, "Maxwell's Treatment of Siemens's Hydraulic Governor," 866; Maxwell, "On Governors," 270, 273.

81. In Fuller's description, Thomson's water-break "regulator consists essentially of a vertical tube with the lower end dipping in water and the upper end branching into two tubular arms.... This forked tube is made to rotate about its vertical axis by being connected to the machine which is to be governed. With water filling the tube and arms, the centrifugal force causes water to be ejected from orifices in the tips of the arms. The work done in raising the water and imparting its exit velocity provides a braking force on the driving machine." Siemens's hydraulic brake similarly uses a "cup-shaped vessel" whose "rotation causes the water in the cup to mount up the sides. When the water reaches the top of the cup it is thrown up by centrifugal force, then checked by surrounding stator blades, and returned to the tank." Both methods made use of positive feedback as a way of addressing "offset" or errors in response, though Maxwell only treats the method described by Siemens. Fuller, "Maxwell's Treatment of Siemens's Hydraulic Governor," 866, 863, 864.

82. Ibid., 875.

83. "On the Vortex Water Wheel" is included as part of a chapter titled "Papers Relating to Fluid Motion" in Thomson, *Collected Papers in Physics and Engineer-ing*, 2–16. This paper was originally published as part of the *Report of the British Association* (Belfast, 1852), while the excerpt describing the patent was published in *Mechanics' Magazine* (January 18, 1851).

84. See Thomson, *Collected Papers in Physics and Engineering*, 12–13.

85. Ibid., 2.

86. Ibid., 2–3. Italics are in the original.

87. Ibid., 15.

88. The attribution of the term *torque* to James Thomson is also given in Fuller, "Maxwell's Glasgow Transcripts," 1611.

89. *A Dictionary of Greek and Roman Antiquities*, ed. William Smith (London: John Murray, Ablemarle Street, 1882), 1140.

167

90. Thomson, *Collected Papers in Physics and Engineering*, 452–457. "On an Integrating Machine having a New Kinematic Principle" was originally published in the *Proceedings of the Royal Society* 24 (1876).

91. Ibid., 454; see also David A. Mindell, *Between Human and Machine: Feedback, Control, and Computing before Cybernetics* (Baltimore: Johns Hopkins University Press, 2002) 38.

92. Thomson, *Collected Papers in Physics and Engineering*, 453.

93. Ibid., 454.

94. These three papers by Kelvin, "An Instrument for Calculating ($\int\phi(x)\psi(x)dx$), the Integral of the Product of Two Given Functions," "Mechanical Integration of Linear Differential Equations of the Second Order with Variable Coefficients," and "Mechanical Integration of the General Linear Differential Equation of Any Order with Variable Coefficients," were published in the *Proceedings of the Royal Society* (1876) and are collected in William Thomson (Lord Kelvin) and Peter Guthrie Tait, *Treatise on Natural Philosophy* (Cambridge: University Press, 1879), 493–504.

95. Thomson, *Collected Papers on Physics and Engineering*, 454.

96. Ibid., 456–457.

97. Ibid., 452n.

98. For Joe Sachs, the standard translations of both *energeia* and *entelecheia* as "actuality" lose what in Aristotle's Greek "pulsates with the dynamic conception of being, and guides a way of seeing the world as organizing itself into every instance of identity it presents." Citing its use of the root *erg-*, he instead translates *energeia* as "being-at-work," and subsequently *entelecheia* as "being-at-work-staying-itself." Joe Sachs, "Introduction," in Aristotle, *Metaphysics*, trans. Joe Sachs (Santa Fe, NM: Green Lion Press, 1999), xl–xli, li–lii.

99. Aristotle, *Nicomachean Ethics*, trans. Christopher Rowe, intro. Sarah Broadie (Oxford: Oxford University Press, 2002), 102 (Bekker, 1098a5ff).

100. Aristotle, *Metaphysics: Book Theta*, trans. Stephen Makin (Oxford: Clarendon Press, 2006), 11 (Bekker, 1050a20ff).

101. Bachelard, *Water and Dreams*, 108.

3 THE AUGMENTATION OF THE INTERFACE

1. Georges Dumézil, *Archaic Roman Religion*, vol. 1, trans. Philip Krapp (1966; Baltimore: Johns Hopkins University Press, 1996), 359, 358.

2. Ibid., 359.

3. Ibid., 361–362.

4. Giorgio Agamben, *Profanations*, trans. Jeff Fort (2005; New York: Zone Books, 2007), 13.

5. Ibid., 12, 14.
6. Giambattista Vico, *On the Most Ancient Wisdom of the Italians*, trans. L. M. Palmer (1710; Ithaca: Cornell University Press, 1988), 96, 102.
7. Ibid., 97. The original *ingenium* has been substituted in place of "wit" as given in this translation.
8. Michael Polanyi, *The Tacit Dimension* (1966; Chicago: University of Chicago Press, 2009), 4, 10. Italics are in the original.
9. Ibid., 33–34.
10. Ibid., 34.
11. Martin Heidegger, *Being and Time*, trans. John Macquarrie and Edward Robinson (New York: Harper and Row, 1962), 98 (original pagination 69).
12. Ibid., 101, 98 (original paginations 71, 69).
13. Ibid., 106 (original pagination 75).
14. Ibid., 107 (original pagination 76).
15. Maurice Merleau-Ponty, *Phenomenology of Perception*, trans. Colin Smith (London: Routledge and Kegan Paul, 1962), 79. The original quote is in Maurice Merleau-Ponty, *Phénoménologie de la perception* (Paris: Éditions Gallimard, 1945), 82.
16. Hubert L. Dreyfus, *What Computers Can't Do: A Critique of Artificial Reason* (New York: Harper and Row, 1972), 204. A discussion of these four areas shapes the concluding section of this work; see ibid., 203–217.
17. Ibid., 206.
18. Ibid., 192.
19. Ibid., xxv; the original citation is Herbert A. Simon and Allen Newell, "Heuristic Problem Solving: The Next Advance in Operations Research," *Operations Research* 6, no. 1 (January-February 1958): 6.
20. Herbert A. Simon, *Models of My Life* (New York: Basic Books, 1991), 206–207.
21. Oliver Selfridge, "Pandemonium: A Paradigm for Learning" (1958), in *Neuro-computing*, ed. James A. Anderson and Edward Rosenfeld (Cambridge: MIT Press, 1988), 117–122.
22. Wendy Hui Kyong Chun, "On Software, or the Persistence of Visual Knowledge," *Grey Room* 18 (Winter 2004): 44.
23. Citing von Neumann as the first to posit the technological singularity, computer scientist and science fiction author Vernor Vinge presents it as a kind of ground zero of futurology: "We are on the edge of a change comparable to the rise of human life on Earth," in which singularity casts "an opaque wall across the future." Technology entrepreneur Ray Kurzweil describes singularity as the moment when "our technology will match and then vastly exceed the refinement and suppleness of what we regard as the best of human traits." Vernor Vinge,

"Technological Singularity," *Whole Earth Review*, no. 81 (Winter 1993): 89–90. Ray Kurzweil, *The Singularity Is Near: When Humans Transcend Biology* (New York: Viking, 2005), 9.

24. Stanislaw Ulam, "John von Neumann 1903–1957," *Bulletin of the American Mathematical Society* 64, no. 3, part 2 (1958): 5. See also Vinge, "Technological Singularity," 90.

25. Anthony G. Oettinger, preface to Dreyfus, *What Computers Can't Do*, xi.

26. Dreyfus, *What Computers Can't Do*, xxxv. Dreyfus relates that Licklider came to his defense at a 1966 RAND symposium after his work was called "sinister," "dishonest," and an "incredible misrepresentation of history."

27. J. C. R. Licklider, "Man-Computer Symbiosis," *IRE Transactions on Human Factors in Electronics* HFE-1 (March 1960): 4, 5.

28. For more on the history of HFE see David Meister, *The History of Human Factors and Ergonomics* (Mahwah, NJ: Lawrence Erlbaum Associates, 1999).

29. Ronald M. Baecker, "A Historical and Intellectual Perspective," *Readings in Human-Computer Interaction: Toward the Year 2000*, ed. Ronald M. Baecker, Jonathan Grundin, William A. S. Buxton, and Saul Greenberg (San Francisco: Morgan Kaufmann, 1995), 41. See also Jonathan Grundin, "A Moving Target: The Evolution of HCI," *The Human-Computer Interaction Handbook: Fundamentals, Evolving Technologies, and Emerging Applications*, ed. Andrew Sears and Julie A. Jacko (New York: Lawrence Erlbaum Associates, 2008), 3.

30. Vannevar Bush, "As We May Think," *Atlantic Monthly* 176, no. 1 (July 1945): 106. A condensed and illustrated version published in *Life* (September 1945) bore the subtitle "A Top US Scientist Foresees a Future World in which Man-Made Machines Will Start to Think." Bush had previously published a number of these ideas: first in a *Technology Review* article titled "The Inscrutable 'Thirties" (1933), where he promoted technologies for augmenting "innovation in knowledge transfer and storage"; and later in a *Fortune* article titled "Mechanization of the Record" (1939). See James M. Nyce and Paul Kahn, "A Machine for the Mind: Vannevar Bush's Memex," in *From Memex to Hypertext: Vannevar Bush and the Mind's Machine*, ed. James M. Nyce and Paul Kahn (San Diego: Academic Press, 1991), 39–66.

31. Alan M. Turing, "Computing Machinery and Intelligence," *Mind: A Quarterly Review of Psychology and Philosophy* 59, no. 236 (October 1950): 434.

32. Douglas Engelbart, "A Conceptual Framework for the Augmentation of Man's Intellect," in *Vistas in Information Handling*, ed. Paul W. Howerton and David C. Weeks (Washington, DC: Spartan Books, 1963), 1. Also quoted in Baecker, "A Historical and Intellectual Perspective," 39.

33. Nicholas Negroponte, *The Architecture Machine: Toward a More Human Environment* (Cambridge: MIT Press, 1970), 9, 11. Also quoted in Baecker, "A Historical and Intellectual Perspective," 43.

34. Stuart K. Card, Thomas P. Moran, and Allen Newell, *The Psychology of Human-Computer Interaction* (Hillsdale, NJ: Lawrence Erlbaum Associates, 1983), 4.

35. Ibid.

36. Knight Dunlap, Letter to the National Research Council, submitted through J. S. Ames, March 21, 1919, 5 pp.; quotes are on pp. 3, 4. Knight Dunlap archives, Archives of the History of American Psychology, University of Akron, Box M570.

37. Franklin V. Taylor, "Psychology and the Design of Machines," *American Psychologist* 12 (1957): 255. Taylor was the namesake of an award granted by the American Psychological Association for "Outstanding Contributions in the Field of Applied Experimental/Engineering Psychology"; in 1957 this award would be granted to J. C. R. Licklider.

38. Lawrence J. Fogel, "A New Concept: The Kinalog Display System," *Human Factors* 1, no. 2 (April 1959): 30–37. I previously treated the Kinalog in Branden Hookway, "Cockpit," in *Cold War Hothouses: Inventing Postwar Culture, from Cockpit to Playboy*, ed. Beatriz Colomina et al. (New York: Princeton Architectural Press, 2004), 49–51.

39. Fogel, "A New Concept: The Kinalog Display System," 30.

40. Ibid., 36, 32, 33.

41. Gilles Deleuze and Félix Guattari, *A Thousand Plateaus: Capitalism and Schizophrenia*, trans. Brian Massumi (1980; Minneapolis: University of Minnesota Press, 1993), 406.

42. Terry Winograd and Fernando Flores, *Understanding Computers and Cognition: A New Foundation for Design* (Norwood, NJ: Ablex Publishing, 1986), 37.

43. Ibid., 164.

44. Ibid., 165, 164.

45. Ibid., 37.

46. Douglas Engelbart, "X-Y Position Indicator for a Display System," U.S. Patent 3,541,541, filed June 21, 1967 and issued November 17, 1970.

47. Ibid.

48. Robert R. Everett, "Whirlwind," in *A History of Computing in the Twentieth Century: A Collection of Essays*, ed. N. Metropolis, J. Howlett, and Gian-Carlo Rota (New York: Academic Press, 1980), 365.

49. Ibid., 375, 376.

50. Kent C. Redmond and Thomas M. Smith, *From Whirlwind to MITRE: The R&D Story of the SAGE Air Defense Computer* (Cambridge: MIT Press, 2000), 81.

51. Ibid., 81–82. The original source is: "Summary Report 7, July 25–October 25, 1950," submitted by the Air Traffic Control Project of the Servomechanism Lab, MIT.

171

52. Interview with William McGill cited in M. Mitchell Waldrop, *The Dream Machine: J. C. R. Licklider and the Revolution that Made Computing Personal* (New York: Viking Penguin, 2001), 108.

53. Licklider, "Man-Computer Symbiosis," 9.

54. Ibid.

55. J. C. R. Licklider, "Graphic Input—A Survey of Techniques," in *Computer Graphics: Utility, Production, Art*, ed. Fred Gruenberger (Washington, DC: Thomson, 1967) 44. Quoted in Waldrop, *The Dream Machine*, 255.

56. Ivan Sutherland, "Sketchpad: A Man-Machine Graphical Communication System," Ph.D. thesis, MIT Department of Electrical Engineering, 1963, 8.

57. Quotes here are drawn from Engelbart, "X-Y Position Indicator for a Display System."

58. Plato, *The Symposium*, in *The Dialogues of Plato*, trans. Benjamin Jowett (New York: Macmillan, 1892), 575.

59. Søren Kierkegaard, *Repetition*, in *Repetition and Philosophical Crumbs*, trans. M. G. Piety (1843, 1844; Oxford: Oxford University Press, 2009), 77, 79.

60. Giorgio Agamben, *Homo Sacer: Sovereign Power and Bare Life*, trans. Daniel Heller-Roazen (Stanford: Stanford University Press, 1998), 16.

61. Kierkegaard, *Repetition*, 78. See also Carl Schmitt, *Political Theology: Four Chapters on the Concept of Sovereignty*, trans. George Schwab (Chicago: University of Chicago Press, 1985), 22; Agamben, *Homo Sacer*, 16.

62. Kierkegaard, *Repetition*, 79.

INDEX

173

175

177